SCARED SH*TLESS

SCARED SH*TLESS

1,003 FACTS

THAT WILL SCARE THE SH*T OUT OF YOU

CARY
McNEAL

A PERIGEE BOOK

A PERIGEE BOOK
Published by the Penguin Group
Penguin Group (USA) Inc.
375 Hudson Street, New York, New York 10014, USA
Penguin Group (Canada), 90 Eglinton Avenue East, Suite 700, Toronto, Ontario M4P 2Y3, Canada (a division of Pearson Penguin Canada Inc.) • Penguin Books Ltd., 80 Strand, London WC2R 0RL, England • Penguin Group Ireland, 25 St. Stephen's Green, Dublin 2, Ireland (a division of Penguin Books Ltd.) • Penguin Group (Australia), 250 Camberwell Road, Camberwell, Victoria 3124, Australia (a division of Pearson Australia Group Pty. Ltd.) • Penguin Books India Pvt. Ltd., 11 Community Centre, Panchsheel Park, New Delhi—110 017, India • Penguin Group (NZ), 67 Apollo Drive, Rosedale, Auckland 0632, New Zealand (a division of Pearson New Zealand Ltd.) • Penguin Books (South Africa) (Pty.) Ltd., 24 Sturdee Avenue, Rosebank, Johannesburg 2196, South Africa
Penguin Books Ltd., Registered Offices: 80 Strand, London WC2R 0RL, England

Text design by Pauline Neuwirth

First edition: September 2012

Library of Congress Cataloging-in-Publication Data

McNeal, Cary.
Scared sh*tless : 1,003 facts that will scare the sh*t out of you / Cary McNeal. —1st ed.
p. cm.
Includes bibliographical references.
ISBN 978-0-399-53782-0
1. Curiosities and wonders. I. Title. II. Title: Scared shitless.
AG243.M3183 2012
031.02—dc23 2012022360

PRINTED IN THE UNITED STATES OF AMERICA

20 19 18 17 16 15 14 13 12 11

CONTENTS

INTRODUCTION

WHAT SCARES YOU?

Death? Crime? Disease? Natural disasters? Or do you suffer from boutique fears like clowns, cannibals, countries whose names end in "-istan" or people who say "nuke-ular" or "supposably" or "conversate"?

Fears vary by individual, but one thing is certain: we're all afraid of something. Put anyone who says otherwise in an airplane plummeting toward earth or in the path of a charging polar bear and see how fearless that person really is. Then help him change into a fresh pair of shorts.

We're all afraid of something.

In this follow-up to the bestselling (thank you!) *1,001 Facts That Will Scare the S*#t Out of You*, I've included another shipload of facts to frighten both the many and the few. Things like bad breath and Celine Dion might not scare you, but they sure as hell scare me, especially if you've just eaten a feta-cheese-and-onion sandwich and start close-talking to me about how much you love Celine's new record. And I don't scare easily.

We're all afraid of something. What's your fear?

Thanks for buying my book. I hope you like it.

CM

CELEBRITY FREAK CLUB

Famous People, Infamous Behavior

SO YOU KNOW HOW PEOPLE like to say that celebrities are just like us? Those people are wrong. Celebs aren't like us. I know it's nice to think you could bond with someone like Angelina Jolie or George Clooney or Pitbull over a beer at T.G.I. Friday's, but you can't. I've worked with plenty of famous folk over the years, and while some of them are lovely people, even the nice ones aren't like you and me. I don't know if it's the adoration or the money or the opportunity to dance competitively on TV, but celebrities are, well, weird. So when they do weird things like shower eighteen times a day or buy property on the moon or eat their roommates, it shouldn't surprise you.

It certainly doesn't surprise me.

FACT 1 ☞ Billionaire Howard Hughes **stored his own urine** in bottles.

FACT 2 ☞ **Hughes also wore empty tissue boxes as shoes. And blew his nose in his socks.**

FACT 3 ☞ Appearing in a tampon commercial at age nineteen, actress Courteney Cox became the first person **to say the word "period"** in that context on American television.

FACT 4 ☞ In 2006, William Shatner was paid $25,000 by an online casino for a **kidney stone** he had recently passed.

FACT 5 ☞ Charlie Chaplin once placed third in a **Charlie Chaplin look-alike contest**.

FACT 6 Actor Charlie Sheen says **he used steroids** to prepare for his role as a pitcher in the movie *Major League*. Of course he did. He's Charlie Sheen.

FACT 7 George Washington has two books from the New York Society Library **still checked out in his name**. Loaned in 1789, the books are more than 222 years overdue, and have accumulated a fine of more than $300,000 in today's dollars.

FACT 8 Oprah Winfrey is chilephobic: **she fears gum chewing**. Her phobia began as a child when she found her grandmother's stash of ABC gum (Already Been Chewed) in a kitchen cabinet.

FACT 9 Alfred Nobel, namesake of the Nobel Peace Prize, **invented dynamite**.

FACT 10 The Gloucestershire, England, airport once used tapes of Tina Turner's music **to drive off birds** from its runways.

FACT 11 **Napoleon Bonaparte was afraid of cats, but he wasn't alone: other ailurophobics include Adolf Hitler, Benito Mussolini and Julius Caesar.**

FACT 12 Early in his career, *Grey's Anatomy* star Patrick Dempsey wanted to **attend clown college and become a juggler**.

FACT 13 **Martha Stewart dated actor Anthony Hopkins**, but dumped him because she couldn't separate him from Hannibal Lecter, the character he portrayed in *The Silence of the Lambs*.

FACT 14 ☞ Italian philosopher and scientist Galileo Galilei's middle finger can be found **on display at Italy's Museo di Storia della Scienza.**

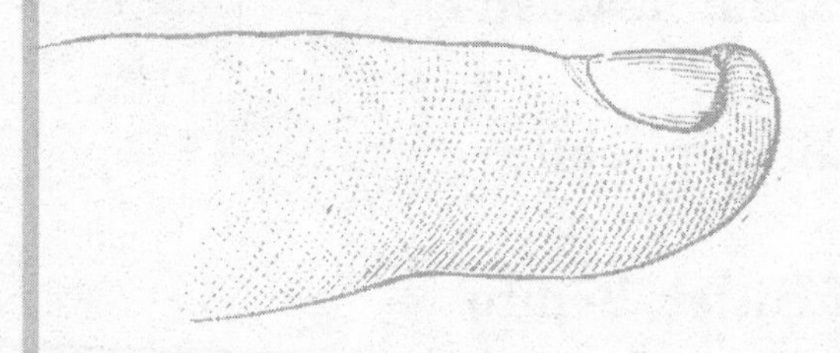

FACT 15 ☞ Actor Billy Bob Thornton fears **flying, bright colors, clowns and antique furniture.**

FACT 16 ☞ Halley's Comet appeared **on the day Mark Twain was born in 1835, and again on the day Twain died in 1910**.

FACT 17 ☞ Rapper Lil Wayne lost his virginity **at age eleven—to a thirteen-year-old girl**.

FACT 18 ☞ Thomas Edison invented the lightbulb in part because he was **afraid of the dark.**

FACT 19 ☞ Mobster Al Capone died in 1947 of **complications from syphilis**.

FACT 20 ☞ Richard Nixon **feared hospitals**; he thought if he ever went into a hospital, he would not come out alive.

FACT 21 ☞ Actress Charlize Theron **was discovered in a bank** when an agent witnessed her throwing a "little tantrum" at a bank teller who refused to cash her check.

THE KING OF WEIRD: ELVIS

Normal has left the building.

FACT 22 Elvis was nominated fourteen times for a Grammy award, **but won only three, all for gospel recordings**. In 1971, he received the Grammy Lifetime Achievement Award.

FACT 23 **Elvis got a C in music** in eighth grade and was told by the teacher than that he had no singing talent.

FACT 24 When Elvis's 1953 song "That's All Right" became popular on Memphis radio, listeners **thought he was black.**

FACT 25 His trademark leg-shaking was due in part to **nervousness**.

FACT 26 ☞ Elvis made his one and only appearance at the Grand Ole Opry in 1954 and tanked, as the conservative crowd found his music and suggestive dancing unsuitable. **He never performed there again.**

FACT 27 ☞ At early concerts, **Presley required police protection** from gangs of teenage boys who resented the singer's popularity with young women and wanted to beat him up.

FACT 28 ☞ A Catholic diocese in Wisconsin complained to FBI director J. Edgar Hoover that "Presley is a definite danger to the security of the United States" and warned that his music would "**rouse the sexual passions of teenaged youth**." Said the church, "That's *our* job."

FACT 29 ☞ The singer's first appearance in Vegas in 1956 was a bust. Conservative, middle-aged audiences didn't like his music, nor did critics; *Newsweek* wrote that Presley's songs were **"like a jug of corn liquor at a champagne party."**

FACT 30 ☞ An early TV appearance on *The Milton Berle Show* in 1956 caused controversy among viewers and critics. One writer proclaimed that popular music "has reached its lowest depths in **the 'grunt and groin' antics of one Elvis Presley**."

FACT 31 ☞ Other TV personalities were similarly unimpressed. **Ed Sullivan declared Elvis "unfit for family viewing"** and Steve Allen considered him "talentless and absurd." Good thing both of them are dead, because that describes about 90 percent of what's on TV now.

FACT 32 ☞ Presley's first appearance on *The Ed Sullivan Show* on September 9, 1956, was seen by approximately 60 million viewers—**a record 82.6 percent of the television audience**.

FACT 33 ☞ After Presley's October 1956 appearance on *The Ed Sullivan Show*, crowds in Nashville and St. Louis **burned him in effigy**.

FACT 34 ☞ In his first full year under contract with RCA, Presley accounted for **more than half of the label's singles sales.**

FACT 35 ☞ After completing his military service, Presley returned to television for the first time as a guest on *The Frank Sinatra Timex Special: Welcome Home Elvis*, even though Sinatra had previously derided rock and roll as "**brutal, ugly, degenerate, vicious . . . sung, played and written . . . by cretinous goons**."

FACT 36 ☞ Elvis was **introduced to drugs** during his army service in Germany; a sergeant gave him amphetamines to help keep Presley awake during lengthy maneuvers.

FACT 37 ☞ Though critically derided, Elvis Presley's films were profitable. Hal Wallis, who produced nine of them, said, "A Presley picture is **the only sure thing in Hollywood**."

FACT 38 Actress Cassandra "Elvira" Peterson met Presley in the 1950s in Las Vegas, where she was working as showgirl. She recalls that Elvis "was so anti-drug when I met him. **I mentioned to him that I smoked marijuana, and he was just appalled**. He said, 'Don't ever do that again.'"

FACT 39 A constant target of death threats since the 1950s, Elvis was known to perform shows with **a pistol in his waistband and another in his boot**.

FACT 40 During one 1973 concert, **four men rushed Elvis's stage** in an apparent attack. The singer used a karate move to fight one off; security stopped the others. The men turned out to be excited fans, not attackers.

FACT 41 Presley **overdosed on barbiturates** twice in 1973, once spending three days in a coma.

FACT 42 ☞ Presley's years of drug abuse were first exposed in the 1977 book *Elvis: What Happened?* written by three of his former bodyguards. The singer **tried to block the book's release** but was unsuccessful.

FACT 43 ☞ By the mid-1970s, Elvis Presley **suffered from multiple ailments**, including glaucoma, high blood pressure, liver damage and an enlarged colon.

FACT 44 ☞ After Elvis's death, his cousin Billy Mann was paid $18,000 **to secretly photograph the corpse**. The picture appeared on the cover of the September 6, 1977, edition of the *National Enquirer*, which remains the tabloid's top-selling issue of all time.

FACT 45 ☞ Graceland was opened to the public in 1982. After the White House, it is the **second most-visited home in the United States,** attracting more than half a million visitors every year.

FACT 46 ☛ During one Elvis appearance on his show, TV host Ed Sullivan noticed a bulge in the singer's crotch and wondered aloud if Presley had **stuffed his pants with a soda bottle**. No, Ed, he was just glad to be there.

FACT 47 ☛ In a bizarre 1970 meeting at the White House, Presley told President Richard Nixon that bands like the Beatles were undermining the country, and **offered his services as an undercover drug agent**.

FACT 48 ☛ Some of Elvis's bejeweled jumpsuits **weighed thirty pounds**. Seems like that would make it hard to jump.

FACT 49 ☛ When Elvis discovered that his wife Priscilla was having an affair with friend Mike Stone, the enraged singer said, **"Mike Stone must die."** But when bodyguard Red West came back to Presley with a price for Stone's contract killing, Presley had calmed. "Let's just leave it for now," he said. "Maybe it's a bit heavy."

FACT 50 One year, Elvis Presley paid **91 percent of his annual income to the IRS.**

FACT 51 Actress Cybill Shepherd dated Elvis in the early 1970s and hinted on *The Oprah Winfrey Show* that **she had to teach the singer how to perform cunnilingus.**

FACT 52 In 1973, Elvis gave Muhammad Ali a boxing robe that read "People's Champion," which Ali wore to fight Ken Norton. When Norton defeated him, **Ali refused to wear the robe ever again.**

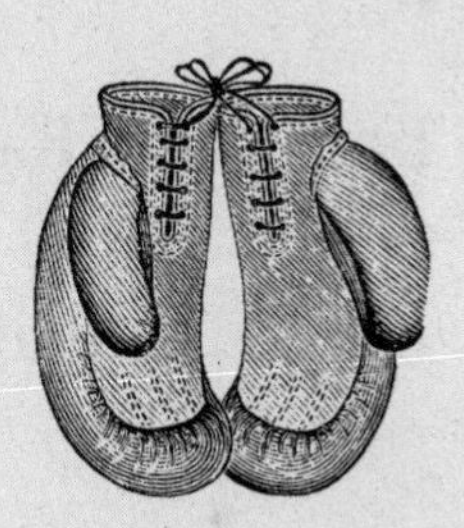

FACT 53 At one point before his death, Elvis consumed an estimated ninety-four thousand calories a day—almost **twice as much as an elephant.**

FACT 54 Sigmund Freud suffered from pteridophobia, a morbid **fear of ferns.**

FACT 55 **Julius Caesar wore a laurel wreath to cover the onset of baldness.** Back then they called it a leaf-over.

FACT 56 President Calvin Coolidge **rarely worked more than four hours a day**. He slept ten hours a night and napped two hours every afternoon.

FACT 57 During a performance at New York's Metropolitan Opera House, tenor Richard Versalle **suffered a fatal heart attack and fell ten feet from a ladder** to the stage just after singing the line "You can only live so long."

FACT 58 ☞ Rapper Big Lurch was convicted in 2003 of the **murder and partial consumption of his roommate** while under the influence of PCP. What, you never heard of roommate and waffles?

FACT 59 ☞ TV's Kelly Osbourne expressed interest in posing nude for *Playboy*, but said that her breasts "would need some airbrushing." *Playboy* founder Hugh Hefner replied, **"We don't airbrush to that extent."**

FACT 60 ☞ Andy Warhol was at one time addicted to the diet pill Obetrol, an **amphetamine** similar to Adderall.

FACT 61 ☞ **French novelist Honoré de Balzac drank so much caffeine that it enlarged his left heart ventricle, which may have contributed to his death.**

FACT 62 ☞ Writer Lewis Carroll **became addicted to laudanum,** an opium-based drug that he took for frequent migranes.

FACT 63 ☞ After a number of automobile accidents, **singer Edith Piaf became addicted to pain pills and alcohol**.

FACT 64 ☞ **In 1939, alarmed by the rise of Nazi Germany, onetime pacifist Albert Einstein urged President Franklin Roosevelt to begin development of an atomic bomb.**

FACT 65 ☞ After his death in 1955, Albert Einstein's **brain was removed and kept in a jar** by Thomas Stoltz Harvey, the pathologist who conducted Einstein's autopsy. Harvey was later fired from his job at Princeton Hospital for refusing to relinquish the organ.

FACT 66 When Greek opera star Maria Callas lost eighty pounds in the early 1950s, one rumor claimed she did so by purposely ingesting a **tapeworm**.

FACT 67 Actor Tim Allen was arrested in 1978 for possession of more than 650 grams of cocaine. He **pled guilty to drug trafficking charges and served two years in prison**.

FACT 68 Mickey Mouse

creator Walt Disney

was **afraid of mice.**

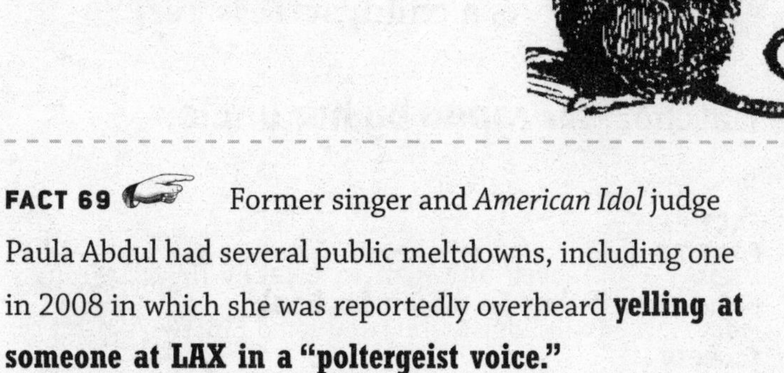

FACT 69 Former singer and *American Idol* judge Paula Abdul had several public meltdowns, including one in 2008 in which she was reportedly overheard **yelling at someone at LAX in a "poltergeist voice."**

FACT 70 When she was fifteen, **actress Charlize Theron's abusive alcoholic father was shot to death by her mother** after he attacked them. Police ruled the act self-defense.

FACT 71 Actress Leighton Meester was **born in a prison hospital** while her mother was serving time on federal drug smuggling charges.

FACT 72 As a child, actress Teri Hatcher was **raped by her uncle.**

FACT 73 Actor Tobey Maguire's father **served time in prison for bank robbery**.

FACT 74 ☞ Oprah Winfrey **had a baby at age fourteen**. The infant died in the hospital several weeks later.

FACT 75 ☞ Singer Nicki Minaj grew up with an abusive father who was an alcoholic and drug addict. Minaj says **she spent her childhood worried** that her dad would kill her mother.

FACT 76 ☞ **Actress Drew Barrymore began drinking at age nine, smoking pot at ten and snorting cocaine at twelve. Barrymore entered rehab at age thirteen, the youngest star ever to do so.**

FACT 77 ☞ Obsessed with cleanliness, Clark Gable showered several times a day and never took a bath because he was disgusted by the thought of **sitting in dirty water**. He also had his sheets changed every single day.

PREMATURE EXTERMINATION: EXAGGERATED REPORTS OF CELEBRITY DEATHS

Chalk it up to wishful thinking.

FACT 78 ☞ After Abe Vigoda was **declared dead by *People* in 1982**, the not-dead actor posed for a retraction photo in the magazine, where he was shown sitting in a coffin, reading the erroneous report.

FACT 79 ☞ In 1987, Vigoda was again **mistakenly reported deceased** during a TV newscast on station WWOR in New Jersey.

FACT 80 ☞ **Rumors of Vigoda's death** follow him to this day, even though the ninety-year-old actor is still alive as of this writing.

FACT 81 ☞ Vigoda had a **real-life brush with death** in 1999 when his American Airlines flight had to make an emergency landing after losing

cabin pressure. Several passengers suffered minor injuries, including the actor. Others reported seeing the Grim Reaper on the tarmac, shaking his fist and saying, "You win this round, Vigoda, but we will meet again."

FACT 82 ☞ The website Abe Vigoda Status (abevigoda.com) maintains an up-to-date report on **whether or not the actor still lives**, as does a Facebook application.

FACT 83 ☞ When rumors began to circulate on Facebook and Twitter in October 2009 that actor **Zach Braff had died**, Braff refuted the rumors in a humorous video on YouTube.

FACT 84 ☞ After Elizabeth Taylor died in 2011 at age seventy-nine, NBC's *Today* show **aired a pre-produced segment on the actress** that claimed she had died at age seventy-seven. The segment had obviously been created before her death and never updated.

FACT 85 In 1973, British music magazine *Melody Maker* ran **a spoof obituary** for rocker Alice Cooper that sent fans into such a frenzy, Cooper made a public announcement to proclaim, "I'm alive, and drunk as usual."

FACT 86 Pop starlet Miley Cyrus **died in a car crash** on September 2008—or so reported a prank story on Digg.com that was picked up by major news outlets and spread quickly across the web.

FACT 87 Two months later, hackers broke into Cyrus's YouTube account and posted a video that claimed—once again—that **the singer had perished**.

FACT 88 ☞ In March 2006, the website iNewswire was fooled into publishing **a phony press release** that claimed comedian Will Ferrell had been killed in a paragliding accident.

FACT 89 ☞ Bob Hope's death was erroneously reported **twice before his actual death** in 2003.

FACT 90 ☞ A 1998 rumor of Hope's death began in—of all places—the U.S. House of Representatives and was **broadcast live on C-SPAN**.

FACT 91 ☞ ABC News falsely declared Sharon Osbourne, wife of rocker Ozzy, dead on its website in 2004. Osbourne had been **fighting cancer for two years**; her obituary had apparently been prepared in anticipation of her death and accidentally posted.

FACT 92 ☞ Web rumors began to circulate in 2005 that TV star Jaleel White (Urkel on *Family Matters*) had **committed suicide** and left a note that read, "Did I do that?"—Urkel's catchphrase.

FACT 93 ☞ When actor Larry Hagman (*Dallas*) was hospitalized for surgery on his transplanted liver in 2004, **rumors began to spread that he was dead**. Hagman debunked the reports in *TV Guide*, saying, "I'm not dead. I'm not retired. I'm simply out of work."

FACT 94 ☞ Actor Will Smith was not killed in a **fall from a cliff in New Zealand** while filming *Men in Black III* in 2011, as some reports claimed.

FACT 95 ☞ *Playboy* founder Hugh Hefner went on Twitter in July 2011 to deny **rumors of his death** from a heart attack.

FACT 96 ☞ Though he is serving time in prison on federal tax charges, actor Wesley Snipes was not **killed in a prison fight** in 2011.

FACT 97 ☞ In December 2011, rocker Jon Bon Jovi shot down rumors of his death by posting a web photo of himself very much alive and holding a sign that read, **"Heaven looks a lot like New Jersey."**

FACT 98 Actor Russell Crowe was **erroneously reported dead** in 2010 by New York City radio station Z-100; his "death" immediately became a trending topic on Google.

FACT 99 After a gaunt Whitney Houston performed at a Michael Jackson tribute concert in New York in 2001, false rumors of her **death from a drug overdose** began to circulate. Houston died in February 2012 from a drug-related drowning.

FACT 100 Teen idol Justin Bieber has been repeatedly rumored dead since 2009, of causes **ranging from suicide to a drug overdose to gunshot wounds** received in a nightclub tiff.

FACT 101 ☞ Pablo Picasso was so poor at times that he **burned his own paintings for warmth.**

FACT 102 ☞ Howard Hughes once made half a billion dollars in **one day**.

FACT 103 ☞ Film critic Roger Ebert and TV titan Oprah Winfrey **dated in the 1980s**. It was Roger who convinced her to syndicate her talk show.

FACT 104 ☞ The FBI called Ted Kaczynski "The Unabomber" because his **early mail bombs** were sent to universities (Un) and airlines (a).

FACT 105 ☞ Just before the Nazis invaded Paris in 1940, authors H. A. and Margret Rey fled the city on bicycles and took with them **the manuscript for *Curious George*.**

FACT 106 ☞ In 1835, John Wilkes Booth's father, Junius, **threatened to kill** President Andrew Jackson.

FACT 107 ☞ A closer look at a photo of Abraham Lincoln giving his **second inaugural address in 1865** reveals among the crowd the faces of Lincoln's assassin, John Wilkes Booth, and accomplices David Herold, George Atzerodt, Lewis Payne, John Surratt and Edmund Spangler.

FACT 108 ☞ **Contrary to legend, George Washington's false teeth were not wooden. Some were gold, ivory or lead; others were horse or donkey teeth.**

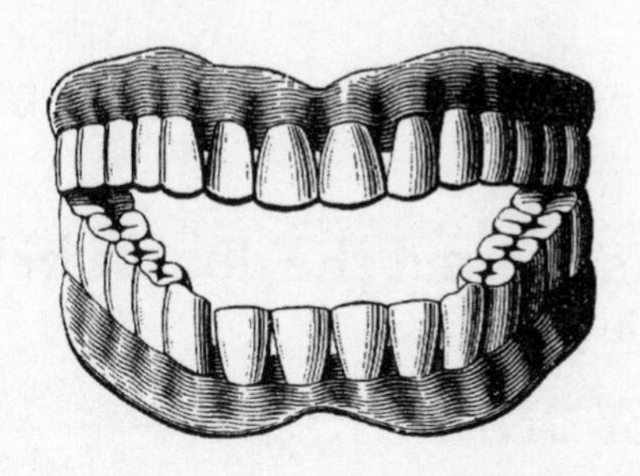

FACT 109 ☞ Inept president **Ulysses S. Grant** was nicknamed "Useless."

FACT 110 ☞ Ineffective president **Dwight D. Eisenhower** was nicknamed "Do-Nothing Ike," "The Golfer," and "World Mangler."

FACT 111 ☞ President Andrew Johnson was **impeached by the House of Representatives** in 1868 but kept his job after the measure failed to pass the Senate by one vote.

FACT 112 ☞ TV game-show host Bob Barker was **taught karate by his friend Chuck Norris**.

FACT 113 ☞ Nobel Prize–winning biologist Francis Crick was **high on LSD** when he discovered the double helix structure of DNA.

FACT 114 ☞ Nicolas Cage **ate a real cockroach** for the film *Vampire's Kiss*. Said Cage, "Every muscle in my body didn't want to do it, but I did it anyway."

FACT 115 ☞ Cage also had **teeth extracted** for his role in *Birdy*.

FACT 116 ☞ Known primarily for his **treachery during the Revolutionary War**, Benedict Arnold was a skilled leader who had shown great valor earlier in the war and saved the Revolution at the Battle of Saratoga.

FACT 117 ☞ Benedict Arnold was **lured into betraying his country** in large part by Peggy Shippen, a teenage daughter of a prominent Loyalist family. Arnold and Shippen later married.

FACT 118 ☞ In 1978, three weeks after marrying his fourth wife, Oscar-winning actor Gig Young apparently **shot her to death** in their New York City apartment and then turned the gun on himself.

FACT 119 Boxing promoter Don King **killed two people.** One murder was ruled justifiable because King was being robbed at the time.

FACT 120 In the other case, King was convicted of second degree murder for stomping a man to death and **spent almost four years in prison**.

FACT 121 Tony Award–winning actor Paul Kelly spent more than two years in prison after **beating another man to death** in 1927.

FACT 122 ☞ Kelly later married the dead man's **widow**.

FACT 123 ☞ Punk rocker Sid Vicious of The Sex Pistols was charged in 1978 with the **stabbing murder of his girlfriend**, Nancy Spungen. Vicious died of a heroin overdose before he could be brought to trial.

FACT 124 ☞ In 2009, famed record producer Phil Spector was convicted of the murder of actress Lana Clarkson. Spector is currently **serving a prison term of nineteen years** to life.

FACT 125 ☞ Rapper C-Murder, aka Corey Miller, was found guilty of murder in 2002 and sentenced to life in prison after **beating and shooting a sixteen-year-old boy** who had allegedly embarrassed Miller during a club's rap contest.

FACT 126 ☞ In 1987, actor Matthew Broderick **killed two women** in Northern Ireland when his car veered into the oncoming lane. Broderick was convicted of careless driving and paid a $175 fine.

FACT 127 ☞ Writer William S. Burroughs shot and killed his common-law wife, Joan Vollmer, in 1951. Burroughs initially claimed he was trying **to shoot a glass off Vollmer's head** and missed.

FACT 128 ☞ Burroughs was **found guilty of manslaughter** and given a two-year suspended jail sentence, serving only two weeks behind bars.

SHOOTING PISTOLS IN THE AIR: CELINE DION

Oh, she's Canadian? That explains it.

FACT 129 ☞ Celine Dion is one of the world's bestselling female singers, with **more than 200 million CDs sold.**

FACT 130 ☞ In 2007, Dion was ranked by *Forbes* as the **fifth richest woman in entertainment**, with an estimated net worth of U.S. $250 million.

FACT 131 ☞ Celine Dion has released **thirty-seven full-length records** in French and English, making her one of only a handful of artists to suck in more than one language.

FACT 132 ☞ Dion is the **bestselling Canadian artist of all time**, edging out Crash Test Dummies, Gino Vannelli and Saga.

FACT 133 In 2003, Dion signed a deal with Coty Inc., to release **Celine Dion Parfums.**

FACT 134 Since its inception, Celine Dion Parfums has grossed more than **$850 million** in retail sales.

FACT 135 **Dion has won numerous accolades from around the world:** Grammy Awards in the United States, Juno and Felix Awards in Canada, and World Music Awards in Europe.

FACT 136 Celine Dion first met future husband René Angélil **when she was twelve years old**. He was sixty-eight. Just kidding, he was thirty-eight. She was still twelve, though.

FACT 137 Dion released her first album at age thirteen after Angélil **mortgaged his home** to finance the recording.

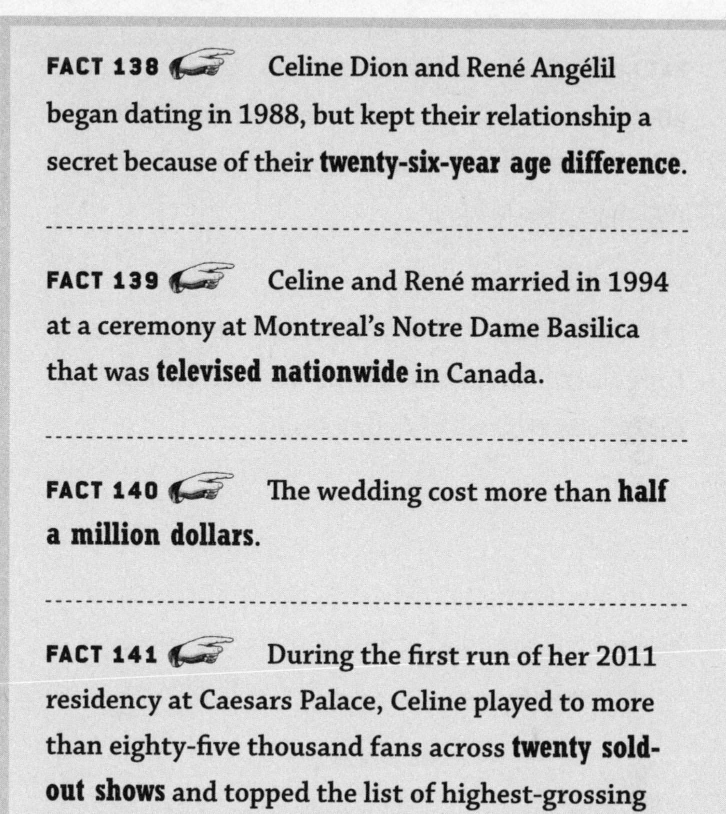

FACT 138 ☞ Celine Dion and René Angélil began dating in 1988, but kept their relationship a secret because of their **twenty-six-year age difference**.

FACT 139 ☞ Celine and René married in 1994 at a ceremony at Montreal's Notre Dame Basilica that was **televised nationwide** in Canada.

FACT 140 ☞ The wedding cost more than **half a million dollars**.

FACT 141 ☞ During the first run of her 2011 residency at Caesars Palace, Celine played to more than eighty-five thousand fans across **twenty sold-out shows** and topped the list of highest-grossing concerts in North America.

FACT 142 ☞ Author Carl Wilson says that **Jamaican gangsters love Celine Dion**. "[A Jamaican friend] told me that when he hears Celine Dion playing in a Kingston neighborhood, he speeds his car up to get out of there, because it means it's a really rough area," he says.

FACT 143 ☞ Says Carl Wilson, "At Jamaican ghetto dances, gangsters scream and **fire pistols into the air** when a Dion song comes on." They're not the only ones.

FACT 144 ☞ In July 2011, attorneys for Celine Dion **forced the closure of the popular Tumblr blog**, Ridiculous Pictures of Celine Dion.

FACT 145 ☞ In 1963, future First Lady Laura Welch (Bush) **ran a stop sign and accidentally crashed her car** into the vehicle of a seventeen-year-old man, killing him. Welch wasn't found to be drinking or speeding, and wasn't charged with a crime.

FACT 146 ☞ Actress Rebecca Gayheart **struck and killed a nine-year-old pedestrian** with her car in Los Angeles in 2001. Gayheart pleaded no contest to vehicular manslaughter and was sentenced to three years' probation and a fine.

FACT 147 ☞ Gayheart was **using her cell phone** at the time of the accident.

FACT 148 ☞ Sharon Osbourne, wife of rocker Ozzy Osbourne, admits to sending her own excrement wrapped in Tiffany boxes to several people who criticized her family. When a journalist criticized her teenage children, Jack and Kelly, Ms. Osbourne **sent a box of excrement** with a note that read, "I heard you've got an eating disorder. Eat this."

FACT 149 ☞ In 1936, eccentric billionaire-to-be **Howard Hughes struck and killed a pedestrian** in Los Angeles. Although Hughes had been drinking earlier that evening, he wasn't drunk at the time of the incident, and charges were dropped.

FACT 150 ☞ **Keith Moon, the late drummer of The Who, accidentally ran over and killed his bodyguard in 1970 while fleeing attackers. Moon wasn't charged.**

FACT 151 ☞ In 1984, Mötley Crüe lead singer Vince Neil was arrested for **drunk driving and vehicular manslaughter** after he crashed into an oncoming car. Neil's passenger was killed, and two occupants of the other car suffered serious injuries.

FACT 152 ☞ Though Neil had a blood-alcohol level more than twice the legal limit at the time of the crash, he was **sentenced to just thirty days in jail**, of which he only served fifteen.

FACT 153 ☞ At age twelve, future Illinois governor and presidential candidate Adlai Stevenson **shot and killed a sixteen-year-old girl** when his gun went off accidentally.

FACT 154 ☞ **Benjamin Franklin almost killed himself while trying to electrocute a turkey.**

FACT 155 ☞ America's nineteenth president, Rutherford B. Hayes, **was nicknamed "His Fraudulency"** by some who believed that Republican power brokers had stolen the election.

FACT 156 ☞ President Warren Harding (1921–23) was nicknamed "Broom Closet Lover" for the **location in which he allegedly liked to rendezvous** with his mistress.

FACT 157 ☞ John Lennon's aunt told him repeatedly as a boy that he would **never make a living playing the guitar**.

FACT 158 ☞ The man who **assassinated Archduke Ferdinand of Austria** in 1914 and sparked World War I, Gavrilo Princip, was only nineteen years old.

THE 99 PERCENT

Troubling Facts About the Rest of Us

PEANUTS CREATOR CHARLES SCHULZ said, "I love mankind; it's people I don't like." I have a friend who puts it more bluntly: "I hate people."

She doesn't mean all people of course. She means stupid people. Weird people. Annoying people. And the stupid, weird, annoying and frightening things they do, like trying to shoot fireworks from their rear end and dying in the process, or dressing up like clowns and keeping people locked in their basements until it's time to eat them.

"Maybe ever'body in the whole damn world is scared of each other," says a character in John Steinbeck's novel *Of Mice and Men*. If they're not, maybe they should be.

FACT 159 ☞ Male drivers involved in fatal car crashes are **twice as likely** as female drivers to be intoxicated.

FACT 160 ☞ Men drink more than women and are responsible for more drunken-driving cases, but **the gap is narrowing quickly**. One reason cited is that women are feeling greater pressures at work and home than before.

FACT 161 ☞ According to the California Office of Traffic Safety, **women accounted for nearly 20 percent of all DUI arrests** in California in 2007, up from 13.5 percent in 1998.

FACT 162 ☞ In July 2009, Diane Shuler drove her minivan the wrong way on Taconic State Parkway in New York and collided with an oncoming vehicle, killing eight people including herself, her daughter, three nieces and three men in the other vehicle. Shuler's autopsy showed that she was **under the influence of alcohol and marijuana** at the time of the accident.

FACT 163 Research has shown that marital trouble is the **most frequent cause of depression** in married men, who can't seem to cope with disagreements as well as women.

FACT 164 **Women apologize more often than men**, but studies suggest that the cause could be a gender difference in what is considered offensive in the first place.

FACT 165 Men lose more money **to Internet fraud** than women: $1.67 to every $1 lost by females.

FACT 166 A UK study found that **men tell lies more often than women.** Men admitted to an average of three lies a day, women only two. But they're all lying, so who can say for sure?

FACT 167 ☞ From 1995 to 2008, eight of ten lightning strike fatalities in the United States were male. The reason is simple: **"Men take more risks in lightning storms,"** says an expert with the National Weather Service.

FACT 168 ☞ Census data in England and Wales reveals that since 1991, for the first time in the twentieth century, **more males than females** are patients at residential mental health facilities in Britain.

FACT 169 ☞ In 2008, four times as many men as women in this country took their own lives, even though **women attempted suicide** three times as often as men.

FACT 170 ☞ A woman attempts suicide **every seventy-eight seconds** in the United States; one is successful every ninety minutes.

FACT 171 ☞ Women are twice as likely as men to have a history of attempted suicide. This is attributed to a **higher rate of mood disorders** among females, such as major depression, dysthymia (chronic mild depression) and seasonal affective disorder.

FACT 172 ☞ **Firearms are now the leading method of suicide for both men and women.**

FACT 173 ☞ Suicide rates for men **rise with age**, most significantly after age sixty-five; rates for women peak between the ages of forty-five and fifty-four.

FACT 174 ☞ Do real men cry? Ninety-nine percent of women say yes, but men see it differently: 39 percent say men should **only cry in response to tragedies**, like the death of a loved one, and 5 percent of males say real men never cry, no matter what.

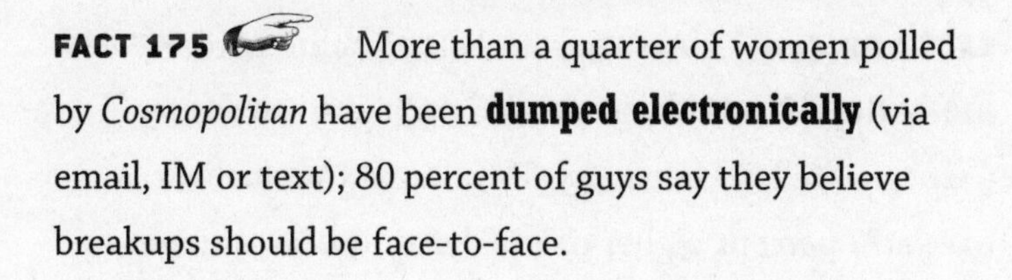

FACT 175 More than a quarter of women polled by *Cosmopolitan* have been **dumped electronically** (via email, IM or text); 80 percent of guys say they believe breakups should be face-to-face.

FACT 176 Depression is the most common women's **mental health problem**.

FACT 177 **Overall rates of psychiatric disorders** are almost identical for both genders, but major depression is twice as common in women.

FACT 178 Doctors are more likely to diagnose depression in women compared with men, **even when they have similar scores** on standardized measures of depression or present with identical symptoms.

FACT 179 **Men are three times more likely** than women to be diagnosed with antisocial personality disorder, and twice as likely to develop an alcohol dependence.

FACT 180 Another stereotype obliterated: **women are better drivers than men**, according to a study by Carnegie Mellon University. They found that male drivers have a 77 percent higher risk of dying in a car accident than women.

FACT 181 During times of food shortages and higher prices—circumstances expected to aggravate with climate change—the **health of females suffers before that of males**. In India, for example, reduced rainfall is more strongly associated with deaths among girls than boys.

FACT 182 **Women are up to fourteen times more likely than men to die from natural disasters.**

FACT 183 ☞ Case studies suggest that public shame, social inhibitions and lack of survival skills contribute to a greater death rate of women in natural disasters. Also, **women often place themselves at higher risk** by caring for children, the sick and the elderly.

FACT 184 ☞ More than 9.4 million women—one in three visitors—**access adult websites** every month.

FACT 185 ☞ A 2003 study by the Harvard School of Public Health found that **mothers pregnant with male babies** ate more than those carrying female babies, but did not gain noticeably more weight than the expecting mothers of girls.

FACT 186 ☞ Men have **more blood** than women: 1.5 gallons versus 0.875 gallons.

PAGING MR. DARWIN: THE STUPID AND THE DEAD

HEY, YOU! Out of the gene pool!

FACT 187 ☞ At a July 2011 protest ride against motorcycle helmet laws in Onondaga, New York, a fifty-five-year-old man **died when he was thrown over the handlebars of his bike** and hit his (helmetless) head on the pavement.

FACT 188 ☞ A Washington man died in 2010 at a racing event when he and a pal **filled a fifty-five-gallon barrel with methanol**, climbed aboard and lit the bunghole on fire, hoping the barrel would shoot across the parking lot like a rocket. It didn't.

FACT 189 ☞ In 2010, after an elevator departed without waiting for him, a handicapped South Korean man began angrily **ramming his wheelchair into the doors**. On the third impact, the doors gave way and the man plunged to his death down the now-empty shaft.

FACT 190 A man fell to his death in the Grand Canyon in 2000 when he ignored warning signs and jumped onto a ledge to collect money tossed there for good luck wishes. He made the first jump, but **the added weight of the coins** made his return jump fall short.

FACT 191 A circus fire-eater in Romania died in 1998 after belching up flammable liquid during his act and catching himself on fire. **No one came to his aid**, as the audience thought the stunt was part of the show.

FACT 192 In 1997, **an Italian stripper suffocated to death** after waiting an hour to jump out of a sealed cake at a bachelor party.

FACT 193 Attila the Hun died on his wedding night in AD 453 after drinking too much, passing out in bed and **drowning on his own blood** from a nosebleed.

FACT 194 ☞ An Austrian man who once had **the world's longest beard** (4.5 feet) died in 1567 when, while fleeing a fire, he stepped on the beard, fell and broke his own neck.

FACT 195 ☞ Danish astronomer Tycho Brahe held his pee so long during a banquet in 1601 (excusing oneself during dinner was considered rude) that his bladder **developed an infection that killed him**. Dying during dinner is also rude.

FACT 196 ☞ In 1911, French tailor Franz Reichelt died while testing his latest invention, a combination overcoat and parachute, by **jumping off the Eiffel Tower**. Reichelt had planned to use a dummy for the test—and, in one sense, did.

FACT 197 ☞ Isadora Duncan, the "mother of modern dance," died in 1927 when one of her trademark scarves got caught in the wheel of a car in which she was riding and **snapped her neck**.

FACT 198 ☞ In 1985, a party for lifeguards at the **New Orleans Recreation Department** to celebrate a drowning-free swimming season came to an abrupt halt when a man was found drowned at the bottom of the pool.

FACT 199 ☞ **A Seattle man died in 2005** after anal intercourse with an Arabian stallion perforated his colon. I wonder if the horse's name was Pokey.

FACT 200 ☞ In 2010, a couple in Sao Paulo, Brazil, was killed while having sex in their car, which they had parked on the shoulder of a busy highway **during morning rush hour**. The car was hit by a large truck whose driver did not see it because of dense fog covering the roadway.

FACT 201 ☞ After the dead naked bodies of a young couple were found next to a Columbia, South Carolina, office building in 2007, police determined that the pair **had fallen while having sex on the pyramid-shaped roof**, where their clothing was found.

FACT 202 The first man **to go over Niagara Falls in a barrel** died fifteen years later after slipping on an orange peel and fracturing his leg, which became infected.

FACT 203 While reviewing troops in 1197, King Henry of France stumbled over **his court dwarf**, causing both men to fall from the balcony to their deaths. It's always the little things that get you.

FACT 204 A day after telling the *New York Times* that **he would live to be a hundred,** Jerome Rodale, founder of the organic food movement, died of a heart attack during an appearance on *The Dick Cavett Show*. Thinking his guest had fallen asleep, Cavett asked, "Are we boring you, Mr. Rodale?"

FACT 205 ☞ An opponent of mandatory seat belt laws died in 2005 when an SUV in which he was traveling slid from an icy highway and rolled several times, **ejecting the unbelted man** from the vehicle.

FACT 206 ☞ In 2007, a UK man being treated for psoriasis burned to death after the paraffin-based cream that had been applied to his body **caught fire as he tried to sneak a cigarette**.

FACT 207 ☞ **Fear of clowns** is called coulrophobia.

FACT 208 ☞ Among the 8 percent of adults **who suffer from phobias,** coulrophobia is common.

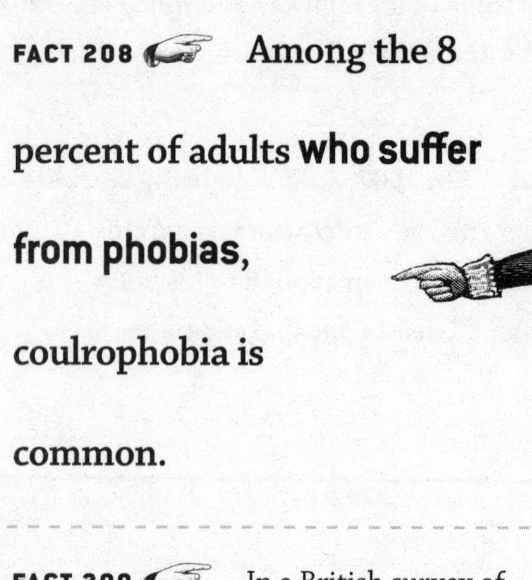

FACT 209 ☞ In a British survey of phobias, **coulrophobia placed third**, outranking common fears such as flying and heights.

FACT 210 ☞ A 2008 study of phobias in England revealed that **children universally fear clowns**, finding them "frightening and unknowable."

FACT 211 Some experts attribute coulrophobia to the **heavy makeup and exaggerated features** of clowns, which can frighten young children.

FACT 212 Horror movie film star Lon Chaney, Sr., once said, "There is nothing laughable **about a clown in the moonlight**."

FACT 213 Many experts point to **obscured facial features** as the most frightening aspect of clowns, and relate this to masked or disfigured movie killers like Michael Myers in *Halloween*, Jason in the *Friday the 13th* movies and Freddy Krueger in the *Nightmare on Elm Street* movies.

FACT 214 Others attribute coulrophobia to the prevalence of **evil clowns in popular media**, such as the child-murdering Pennywise in Stephen King's *It* and the clown doll that attacks a boy in *Poltergeist*.

FACT 215 Fear of **Santa Claus** is a type of coulrophobia.

FACT 216 One of the **worst serial killers in U.S. history**, John Wayne Gacy entertained kids **dressed as a clown**.

FACT 217 Gacy was convicted of sexually assaulting and killing **thirty-three boys** and young men.

FACT 218 Paul Kelly, son of famous clown Emmett Kelly Jr., was arrested in 1978 for the murders of two of his homosexual lovers. Kelly admitted to the slayings, but **listed his clown alter ego Willie as an accomplice.** And I'm sure the cops totally bought it. I bet there's still an APB out on Willie the Clown thirty-five years later.

FACT 219 **Coulrophobia may result** from the incongruity of the exaggerated expressions of joy on the faces of clowns and their aggressive, mischievous behavior.

FACT 220 Some health activists want McDonald's to **drop Ronald McDonald** as a mascot because he markets unhealthy food to children.

FACT 221 Coulrophobia can be treated with **exposure therapy**, which presents patients with photos and dolls of clowns to help them slowly work through their fears.

FACT 222 Famous coulrophobics include **Johnny Depp, Daniel Radcliffe, Billy Bob Thornton and Sean "P. Diddy" Combs**.

FACT 223 Coulrophobic Sean Combs has been known to include **a "no clowns" clause** in contracts—even at the risk of being banned from his own shows.

FACT 224 Johnny Depp attributes his coulrophobia to **childhood nightmares about clowns** leering at him.

FACT 225 Planking (lying down in an unusual place) began in 2006 when two Britons created a Facebook page that attracted **fifteen hundred followers within the first two weeks**, and now boasts more than nineteen thousand image submissions.

FACT 226 **Planking (aka the Lying Down Game)** has spread to the rest of the world and goes by many names, including "playing dead," "à plat ventre" ("on one's belly"), "extreme lying down" and "facedowns."

FACT 227 The **official Facebook page for planking** had more than 130,000 fans in the first week, and currently boasts more than 715,000.

FACT 228 Planking's origin might be in the 1993 movie *The Program*; its trailer features a scene of a **football player proving his bravery** by lying down in the middle of a highway as cars fly by.

FACT 229 After several teens were killed while **mimicking a lying-down** scene from 1993's *The Program*, the scene was cut from the movie.

FACT 230 **According to the official Facebook page, the rules of planking include lying facedown and expressionless, keeping legs straight with toes pointed, having arms by side with fingers pointed and boldly declaring that you are, in fact, planking.**

FACT 231 ☞ **Planking began to grow in popularity in March 2011 after Australian rugby player David "Wolfman" Williams planked during a match.**

FACT 232 ☞ In August 2011, *Playboy* Playmate Anna Sophia Berglund posted on her Twitter page, "I got Hef to plank! @playboy" and included a link to a photo of ***Playboy* founder Hugh Hefner planking** on a table in what appears to be the Playboy Mansion.

FACT 233 ☞ The fad made news in September 2009, when seven doctors and nurses working at the Great Western Hospital in Swindon, England, were **suspended for planking while on duty**.

FACT 234 ☞ Some critics of planking, including rapper Xzibit and author Marcus Rediker, call the activity racist, claiming that it is **reminiscent of how Africans were stacked in slave ship hulls** for transport to America.

FACT 235 ☞ Supporters of planking **deny any intentional connection** between the activity and slave ships.

FACT 236 ☞ Planking has been **criticized for being dangerous**, with people posing on ledges of tall buildings, on railroad tracks and in high trees to compete with other plankers.

FACT 237 ☞ **In May 2011, a twenty-year-old Australian man fell seven stories to his death while attempting to plank between two balcony railings of an apartment building.**

FACT 238 ☞ Police in Australia have warned that people caught planking on private property will be charged with trespassing or the more serious charge of **unauthorized high-risk activity**.

FACT 239 ☛ After a photo of his son planking appeared on the front page of the *New Zealand Herald* in May 2011, **Prime Minister John Key defended the practice** and said his son learned about it from him.

FACT 240 ☛ In 2011, American journalist Michelle McMurray **declared First Annual Global Planking Day on May 25** and encouraged readers to celebrate the inaugural holiday.

FACT 241 ☛ Planking has spawned **several copycat fads**, including "teapotting" (bending the arms in a shape of a teapot in reference to the children's song "I'm a Little Teapot"), "owling" (squatting like an owl) and "dogging" (putting your legs over your arms and dragging your butt along the ground after you poop).

SERIAL KILLERS: GLUTTONS FOR PUNISHMENT

Come on, guys, save some for the rest of us.

FACT 242 ☞ Nicknamed the Sadistic Aristocrat, Gilles de Rais was a fifteenth-century French nobleman who liked to **rape, torture and murder** young peasant boys. Yep, that's sadistic, all right.

FACT 243 ☞ **America's first serial killers** might have been brothers Micajah and Wiley Harpe, who killed at least forty men, women and children in the late 1700s.

FACT 244 ☞ A vigilante mob hunted down killer Micajah Harpe in 1799 and **placed his severed head on a pike** at a crossroads in western Kentucky still known as "Harpe's Head."

FACT 245 ☞ Wiley Harpe was **captured and hanged in 1804**. He got to keep his head.

FACT 246 England's Mary Ann Cotton is believed to have **killed twenty-one people**, including her own mother, several of her children, four husbands and two lovers, in the late 1800s.

FACT 247 From 1884 to 1885, **the Austin Axe Murderer** killed at least seven women, most of them servants, in Austin, Texas.

FACT 248 Victims of the Austin Axe Murderer were slashed or axed to death; some had **spikes driven through their faces** or ears.

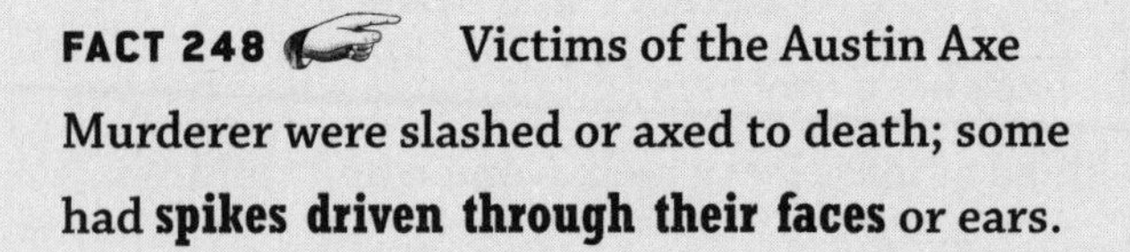

FACT 249 The Austin Axe Murderer **was never caught**. He must have split. Ha! Get it?

FACT 250 The Bloody Benders were a family of four immigrant innkeepers in 1870s Kansas who worked as a team to **kill and rob at least eleven guests**, including women and children.

FACT 251 Italian serial killer Leonarda Cianciulli came to be known as "The Soap-Maker of Correggio" after disposing of her victims' bodies by **turning them into soap**. She was the world's first green mass-murderer.

FACT 252 Joseph "The French Ripper" Vacher murdered, raped and mutilated at least **eleven people in France** over a three-year period beginning in 1894.

FACT 253 1920s serial killer Carl Panzram, who **called himself "rage personified,"** confessed to twenty-one murders and the rapes of more than a thousand young boys.

FACT 254 About his crimes, serial killer Carl Panzram said, **"I am not the least bit sorry."** If nothing else, you have to give him points for honesty.

FACT 255 ☞ French serial killer Henri Landru preyed on lonely widows during the First World War by seducing them, taking their money, then killing them and **cooking their bodies in his oven**.

FACT 256 ☞ Hélène Jegado was a domestic servant-turned-nun in nineteenth-century France who killed as many as thirty-six people by **arsenic poisoning**, including her own sister.

FACT 257 ☞ After serving time for poisoning several patients in Illinois, Dr. Thomas Cream moved to London and resumed his murderous ways until he was caught and **sentenced to death in 1892**. His hangman claims that Cream confessed to being Jack the Ripper. What Cream really said was, "I hate my fucking name."

FACT 258 ☞ English physician William Palmer, aka "Palmer the Poisoner," was convicted of one murder but suspected of many others, including those of **his wife, brother, four infant children and mother-in-law**.

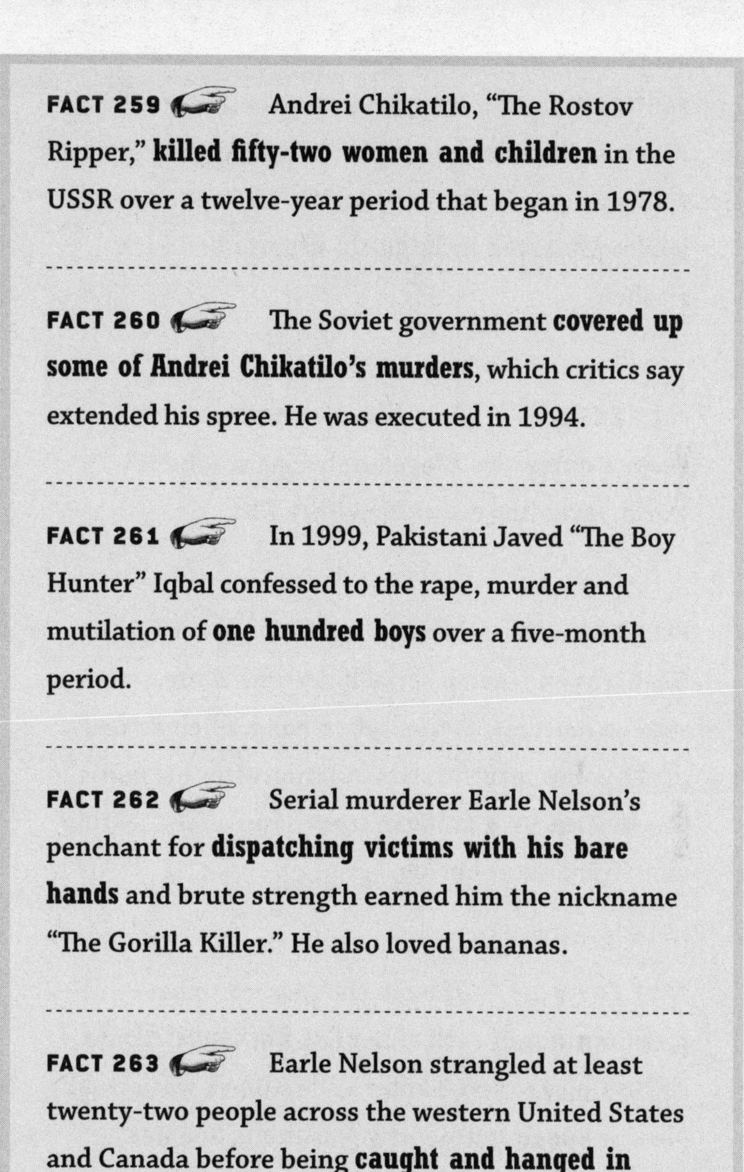

FACT 259 ☞ Andrei Chikatilo, "The Rostov Ripper," **killed fifty-two women and children** in the USSR over a twelve-year period that began in 1978.

FACT 260 ☞ The Soviet government **covered up some of Andrei Chikatilo's murders**, which critics say extended his spree. He was executed in 1994.

FACT 261 ☞ In 1999, Pakistani Javed "The Boy Hunter" Iqbal confessed to the rape, murder and mutilation of **one hundred boys** over a five-month period.

FACT 262 ☞ Serial murderer Earle Nelson's penchant for **dispatching victims with his bare hands** and brute strength earned him the nickname "The Gorilla Killer." He also loved bananas.

FACT 263 ☞ Earle Nelson strangled at least twenty-two people across the western United States and Canada before being **caught and hanged in 1928.**

FACT 264 ☞ Hungarian serial murderer Béla Kiss is believed to have killed at least twenty-four young women in the early 1900s. He kept their bodies **preserved in large tin drums** filled with alcohol.

FACT 265 ☞ **Béla Kiss was never caught** despite numerous alleged sightings around the world, including one in New York City.

FACT 266 ☞ Mohammed "The Desert Vampire" Bijeh was an Iranian serial killer who confessed to sixteen murders. Sentenced to hang, Bijeh actually died by slow, painful strangulation after his noose was **hoisted by a crane** in front of an angry, jeering mob. Iran: never boring.

FACT 267 ☞ Though she claimed to have premonitions of each of her **five husbands' deaths,** 1930s Chicago serial killer Tillie Klimek was actually planning their murders by poisoning. She was successful with four of the five.

FACT 268 ☞ German Bruno Ludke **killed at least eighty people**, most of them women, over a fifteen-year spree that began in 1928.

FACT 269 ☞ Declared insane (ya think?), serial killer Bruno Ludke was sent to a Vienna hospital, where he was subjected to **numerous experiments** before being executed in 1944.

FACT 270 ☞ "Lonely Hearts Killers" Raymond Fernandez and Martha Beck **seduced, robbed and killed vulnerable women** in the 1940s before being caught and executed in 1951.

FACT 271 ☞ In 1944, the false promise of **safe passage from Nazi-occupied France** led at least sixty-three Jews to their deaths at the hands of serial killer Marcel Petoit.

FACT 272 ☞ Notorious UK murderers Fred and Rose West **took the lives of thirteen women and girls**, including Fred's first wife and two of his daughters, before their capture in 1992.

FACT 273 ☞ Miyuki Ishikawa, nicknamed "Oni-Sanba" (demon midwife), was responsible for the deaths from neglect of at least **103 Japanese infants** in her care.

FACT 274 ☞ South Korean policeman Woo Bum-kon is responsible for one of the largest killing sprees in modern times, murdering fifty-six people and wounding thirty-five more in an **eight-hour rampage** in 1982.

FACT 275 ☞ After his conviction for fifty-two murders, Ukranian Anatoly "The Terminator" Onoprienko issued a press release from his prison cell to say that he had **hoped to achieve the world record** for killing.

FACT 276 ☞ Nicknamed **"The Monster of the Andes,"** Pedro Alonso Lopez is thought to have murdered as many as three hundred young girls in Colombia, Peru and Ecuador in the late 1970s and early '80s.

FACT 277 Mass murderer Pedro Alonso Lopez served eighteen years of a life sentence in Ecuador before being deported in 1998 to Colombia, where he promptly disappeared. **His whereabouts are unknown** to this day.

FACT 278 **Australia's two most prolific convicted serial killers** are James Miller, who raped and murdered seven women in the 1970s, and Ivan Milat, the notorious "backpack" murderer, who killed seven tourists in the 1990s.

FACT 279 The 2005 horror film *Wolf Creek* is based in part on **Ivan Milat's murders**.

FACT 280 Pietro Pacciani, "The Monster of Florence," **was convicted in 1994** of murdering eight couples in lovers' lanes between 1968 and 1985.

FACT 281 ☞ In his book *The Silence of the Lambs*, author Thomas Harris based fictional killer Jame Gumb, aka Buffalo Bill, on as many as **six real-life serial murderers**, including Ed Gein, Ted Bundy, Gary Heidnik, Jerry Brudos, Edmund Kemper and Gary Ridgway.

FACT 282 ☞ Henry Lee Lucas was considered **one of America's most prolific killers** after he confessed to murdering more than six hundred people, a figure police later determined was highly exaggerated.

FACT 283 ☞ Despite **confessing to hundreds of murders he didn't commit**, Henry Lee Lucas was convicted of killing eleven people, including his own mother.

FACT 284 ☞ After the arrest of German serial killer Karl Denke in 1924, police found the butchered remains of his nearly three dozen victims **pickled in jars in his basement**.

FACT 285 ☞ Denke told police that he had **eaten nothing but human flesh** for three years before his capture, and had sold leftover meat to his neighbors as "boneless pork."

FACT 286 ☞ Albert Fish was a kindly grandfather figure who confessed to **molesting more than four hundred children**, many of whom he tortured, killed and ate.

FACT 287 ☞ After Albert Fish murdered and ate ten-year-old Grace Budd in 1934, **he wrote to her mother** to describe how delicious the child tasted.

FACT 288 ☞ Albert Fish was **put to death by electrocution** in 1936 at New York's Sing Sing prison.

FACT 289 ☞ Former Liberian warlord Joshua Blahyi confessed to **human sacrifices and cannibalism of children** during his conflict with President Charles Taylor's militia.

FACT 290 ☞ In 2008, a group of illegal Dominican immigrants en route to Puerto Rico **resorted to cannibalism** after they were lost at sea for over two weeks. No one thought to pack a fishing pole?

FACT 291 ☞ Cannibal killer Dorangel Vargas, "The Hannibal Lecter of the Andes," told the press that he **preferred the taste of men to women,** and never ate hands, feet or testicles. "I have standards, you know," said Vargas.

FACT 292 Vargas also eschewed the elderly, he said, because their flesh was **"contaminated and very tough."**

FACT 293 Armin "The Cannibal of Rotenburg" Meiwes filmed himself killing and eating computer engineer Bernd Brandes, forty-two, whom he had met after posting messages in Internet chat rooms **seeking "men for slaughter."**

FACT 294 In 2008, twenty-two-year-old Canadian Tim McLean was **stabbed, beheaded and cannibalized** by the man sitting next to him on a Greyhound Canada bus.

FACT 295 The Korowai of Papua New Guinea are one of very few tribes still believed to **eat human flesh as a cultural practice**.

FACT 296 ☞ Anthropophagy (cannibalism) **is not illegal in most countries** or even most states in the United States. People who eat human flesh are usually charged with other crimes, such as murder or desecration of a body.

FACT 297 ☞ Some settlers of colonial Jamestown resorted to cannibalism during a period known as the Starving Time (1609–10), **digging up corpses** when food supplies ran out.

FACT 298 ☞ Survivors of the whaling ship *Essex*, which was sunk by a sperm whale in the Pacific Ocean in 1820, **resorted to cannibalism** during three months adrift at sea. By the time they were rescued, only two survivors remained.

FACT 299 ☞ After Alexander Pearce led a **group escape from Macquarie Harbour Penal Settlement** in Tasmania in 1922, his men killed and cannibalized one another for weeks until Pearce was caught, the last escapee alive.

FACT 300 ☞ There were **reports of cannibalism** during the Great Chinese Famine of 1958–61, during the famine of 1571 in Russia and Lithuania, and in World War II in Nazi concentration camps and during the Siege of Leningrad.

FACT 301 ☞ In 1947, a war crimes tribunal prosecuted **thirty Japanese soldiers** who had killed and consumed nine American airmen in Chichijima in 1945. Five Japanese officers were hanged.

FACT 302 ☞ **According to author James Bradley, in his book *Flyboys: A True Story of Courage*, some of the Chichijima airmen were eaten for sustenance over several days, their Japanese captors amputating limbs only as needed to keep the meat fresh.**

FACT 303 In 1972, a Uruguayan plane carrying forty-five passengers, mostly students and rugby players, crashed in the Andes Mountains. With no food and bad weather preventing their rescue, the thirty-three survivors were **forced to feed on the bodies of the dead** to survive.

FACT 304 **America's most famous cannibals, the Donner Party, would have survived their journey had they not taken a so-called shortcut around the Great Salt Lake that cost them valuable time. Though weeks behind, they still only missed making it through Donner Pass before heavy snows by one day.**

EAT, DRINK AND BE SCARY: FOOD TO FEAR

Deliciousness kills, but you'll die happy.

FACT 305 ☞ There is **more real lemon juice** in Lemon Pledge furniture polish than in Country Time Lemonade, even though they taste the same.

FACT 306 ☞ According to USFDA standards, a cup of orange juice is allowed to contain **ten fruit fly eggs and two maggots**. But don't worry, they don't drink much.

FACT 307 ☞ **McDonald's McRib sandwich** contains some of the same ingredients used to manufacture gym mats and running shoes—two things McRib eaters know nothing about, so they won't notice.

FACT 308 ☞ Some McDonald's **salads contain more fat** than their burgers.

FACT 309 ☞ More than **2 billion pounds** of bacon are produced every year in the United States.

FACT 310 ☞ If the average piece of bacon is one ounce, Americans consumed **32 billion pieces of bacon** in 2009 alone.

FACT 311 ☞ The USDA defines bacon as **"the cured belly of a swine carcass."** Never has anything that sounds so disgusting tasted so good.

FACT 312 ☞ Chef Anthony Bourdain **has called bacon the "gateway protein"** for its ability to turn vegetarians back into meat eaters.

FACT 313 ☞ Sixty-eight percent of bacon's calories come from fat, almost half of which is **saturated.**

FACT 314 ☞ Every ounce of bacon contains **thirty milligrams of cholesterol.**

FACT 315 ☞ Eating foods **rich in saturated fats like bacon** can raise your cholesterol levels, increasing your risk of heart disease and stroke.

FACT 316 ☞ Researchers have concluded that regular consumption of processed meats (bacon, ham, hot dogs, sausage and more) can lead to **higher risk for prostate cancer** and several other cancers.

FACT 317 ☞ The cancer risk of processed meats could have to do with nitrates, which are typically used as preservatives in processed meat and **morph into cancer-promoting N-nitroso compounds** when digested.

FACT 318 ☞ The **high content of nitrates** may also contribute to COPD (chronic obstructive pulmonary disease), which includes emphysema and chronic bronchitis. In COPD, changes occur in your lungs that make breathing difficult. But even with their final breaths, many people will ask for bacon.

FACT 319 ☞ COPD is the **third leading cause of death** in America, claiming the lives of more than 124,000 Americans in 2007.

FACT 320 ☞ **Risk of cancer from processed meats** like bacon could also be related to carcinogenic PAH (polycyclic aromatic hydrocarbon) compounds, which can be produced during processing.

FACT 321 ☞ A 2010 study by the Harvard School of Public Health **linked processed meats with a 42 percent higher risk of heart attacks** and 19 percent higher risk of diabetes. The same risk was not found in consuming unprocessed red meats.

FACT 322 Scientists have confirmed that **fattening foods like bacon and donuts may be addictive.** A 2010 study suggests that high-fat, high-calorie foods affect the brain in a manner similar to cocaine and heroin, overloading pleasure centers and requiring increasing amounts of the drug or food.

FACT 323 ☞ The FBI estimates that more than **half a million pedophiles** are online every day.

FACT 324 ☞ Pedophiles regularly create **bogus online profiles on social networking sites** to stalk victims and share information with other pedophiles.

FACT 325 ☞ Roughly 40 percent of investigations worked under the FBI's Cyber Division in 2007 were related to **child pornography and child sexual exploitation**.

FACT 326 ☞ More than half of adolescents and teens have been **bullied online**. Cyber-bullying can lead to anxiety, depression and suicide.

FACT 327 ☞ **A third of young people have been threatened online;** **a third have been bullied repeatedly via the web or cell phone texting.**

FACT 328 ☞ About one in five teens has posted or sent **sexually suggestive or nude pictures of themselves** to others online.

FACT 329 ☞ A 2009 report from the U.S. Department of Justice found that nearly a quarter of stalking victims also experienced **some form of cyber-stalking**; some were electronically monitored with spyware, bugging or video surveillance.

FACT 330 ☞ One in every ten American consumers has been victimized by identity theft. There were **10 million victims of identity theft** in the United States in 2008 alone. They're the ones walking around saying, "Who am I?"

FACT 331 ☞ Households with yearly incomes higher than $70,000 are **twice as likely** to experience identity theft as those with incomes under $50,000.

FACT 332 ☞ Up to 18 percent of identity theft victims **don't learn about the crime for four years** or more.

FACT 333 ☞ The average victim spends **330 hours and as much as $1400** repairing the damage caused by identity theft.

FACT 334 ☞ Seventy percent of identity theft victims have **difficulty removing negative information** from their credit reports afterward.

FACT 335 ☞ Forty-three percent of identity theft victims **know the perpetrator**. Unfortunately, many don't *know* they know the perpetrator, or they would let the perpetrator know they know by kicking his perpetrating ass.

FACT 336 ☞ **Internet addiction** is defined as six hours a day of nonessential Internet use for three months or more.

FACT 337 ☞ **Every day, 9 to 15 million people in the United States use the Internet. The amount of use goes up by 25 percent every three months.**

FACT 338 ☞ In South Korea, 11 percent of school-aged youth are considered at high risk for **Internet addiction**.

FACT 339 In China, **96 percent of teenagers use instant messaging**; 10 percent of them can be classified as messaging addicts.

FACT 340 Studies have shown that the more time a person spends on the Internet (excluding work-related usage), the higher his or her risk for **social problems and self-esteem issues**.

FACT 341 **Adolescents who play more than an hour of video games a day have a much greater risk for ADHD and more intense symptoms of ADHD than other kids.**

FACT 342 In 2003, the SQL Slammer worm **infected 90 percent of the world's unprotected computers in ten minutes** and significantly slowed traffic across the Internet.

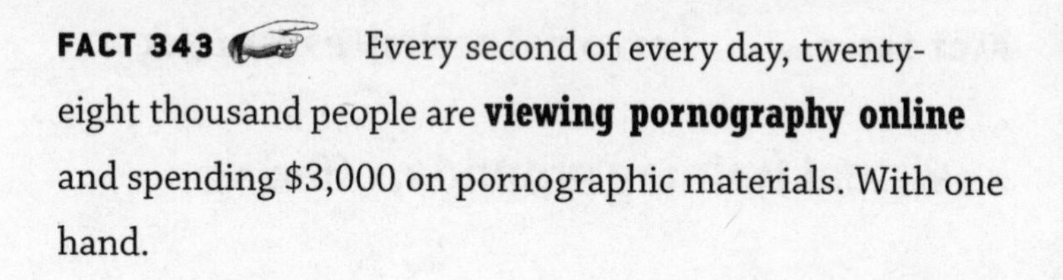

FACT 343 Every second of every day, twenty-eight thousand people are **viewing pornography online** and spending $3,000 on pornographic materials. With one hand.

FACT 344 A new pornographic video is produced in the United States **every thirty-nine minutes**—and it shows.

FACT 345 Child pornography is a **$3 billion annual industry** and one of the fastest-growing online businesses.

FACT 346 In 2008, the Internet Watch Foundation identified **1,536 individual child abuse domains**; 58 percent of them were hosted in the United States.

FACT 347 **The number of terrorist websites** has increased exponentially over the last decade, from fewer than one hundred in the late 1990s to more than forty-eight hundred in 2007.

FACT 348 ☞ **Terrorist websites can serve as virtual training grounds, offering tutorials on building bombs, firing surface-to-air missiles, shooting at U.S. soldiers and sneaking into Iraq from abroad.**

FACT 349 ☞ **Justin Bieber** currently has almost 24 million followers on Twitter, more than Barack Obama (16 million) and the Dalai Lama (4.5 million).

FACT 350 ☞ Scareware, which is fraudulent and often malicious software sold via pop-up warnings of computer viruses, is one of the **fastest-growing and most prevalent** types of Internet fraud.

FACT 351 ☞ Security-software firm McAfee saw a **400 percent increase in scareware incidents** reported in 2009 and predicts the use of scareware will be the most costly type of online scam in years to come.

FACT 352 ☞ A 2010 study by Google found eleven thousand domains hosting **fake antivirus software**, accounting for half of all malware delivered via Internet advertising.

FACT 353 ☞ Facebook hosts a staggering 140 billion photos: that's ten thousand times more photos than the Library of Congress and roughly **4 percent of all the photos ever taken**.

FACT 354 ☞ Facebook's oldest user is a **103-year-old English grandmother** who updates her page using an iPad. She logs on daily to see if she's still alive.

FACT 355 ☞ The **deadliest fire in American history** occurred in Peshtigo, Wisconsin, on October 8, 1871, the same day as the Great Chicago Fire, and killed ten times as many people.

FACT 356 ☞ Rebecca Black **received multiple death threats** over her "Friday" song and video, after it was dubbed "the worst video ever made."

FACT 357 ☞ The Nazis invented an **exploding chocolate bar**, though there is no record of it ever being used.

FACT 358 ☞ **The Nazis also pioneered the concept of the "shoe bomb" decades before "Shoe Bomber" Richard Reid was caught trying to blow up an airplane in 2001.**

FACT 359 ☞ A teenager won first prize in a Virginia science competition after demonstrating that **laboratory mice subjected to hours of heavy metal music** became so angry and aggressive that they killed one another.

FACT 360 ☞ **Repeated meals of bad spaghetti** caused a 1950 riot by prisoners at Alcatraz. It must have been pretty awful. How do you ruin spaghetti? Even *I* can make that.

FACT 361 Disruptive inmates at Los Angeles' Men's Central Jail are fed a "disciplinary loaf"—**an entire meal molded into a baked log** that is nutritious but unpleasant.

FACT 362 At least one American company can turn your remains into **a synthetic diamond** to be worn by a loved one after you die.

FACT 363 Want to go out with a bang? Several American companies will **pack a portion of your ashes into professional-grade fireworks** and stage a memorial display for your survivors.

FACT 364 **One in ten managers** who died on the job in 2010 was murdered.

FACT 365 In 2010, more on-the-job deaths occurred in "**transportation and material-moving occupations**" than in any other category.

FACT 366 ☞ Two hundred and twenty-four people died at work in 2010 after being "**caught in or compressed by equipment or objects**."

FACT 367 ☞ In 2010, 258 people **killed themselves** while at work.

FACT 368 ☞ The IRS is immediately notified anytime you or anyone else withdraws **$10,000 or more** from the bank.

OH, THE PLACES YOU'LL GO... IN YOUR PANTS

The World's Scariest Locations

I WAS TALKING TO a buddy of mine a few weeks back. He had just returned from a family vacation at Disney World in Florida.

"How was it?" I asked him.

"Scary."

Scary? The Magic Kingdom? The happiest place on Earth? The land of wishes and dreams?

"Yeah," he said. "Scary."

I've been to Disney and I get what he meant—sort of—but he was speaking in hyperbole. You want scary? I mean, truly scary? Go to the Aokigahara Suicide Forest in Japan. Check out Lake Nyos in Cameroon or the Chapel of Bones in Portugal. Spend some time in South Africa or Juàrez, Mexico or the Sudan, and you'll get a clearer picture of scary. A place like Disney might be crowded and expensive, but at least you can spend time there without being freaked the F out. Or dying.

FACT 369 Among the twenty thousand anatomical oddities on display at the Mütter Museum in Philadelphia are a **nine-foot-long human colon** and a slice of someone's face suspended in fluid.

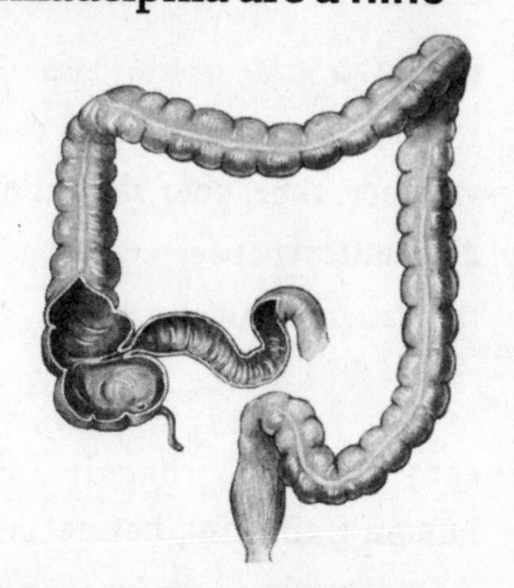

FACT 370 Since the 1950s **more than five hundred people have taken their own lives** at the Aokigahara Suicide Forest near Mount Fuji in Japan, which is littered with bones, makeshift nooses and flowers left by grieving friends and family.

FACT 371 The dangerous border town of Ciudad Juárez, Mexico, saw **2,600 deaths** from drug-related violence in 2009.

FACT 372 Grozny, Chechnya, was named **"Most Destroyed City on Earth"** by the United Nations in 2003. Most residents fled long ago; those who remain are under constant threat from local mafia and gangs.

FACT 373 Kinshasa, the capital and largest city of the Democratic Republic of the Congo, is **plagued by gang violence, rape, poor sanitation, disease and constant civil unrest** between warring factions trying to control the nation's diamond mines.

FACT 374 Karachi, Pakistan, is widely known as a **human trafficking hotspot** and the primary gateway for smuggling sex slaves into the West.

FACT 375 With a yearly average of 1,220 violent crimes per 100,000 residents, Detroit, Michigan, recently topped *Forbes* magazine's list of **most dangerous cities in America**.

FACT 376 Caracas, Venezuela, was dubbed **"Murder Capital of the World"** by *Foreign Policy* magazine in 2008.

ALASKA: IT'S ONLY COLD UNTIL YOU DIE

Meet the state that's a lot like my sister-in-law: huge, icy and just looking for a reason to kill you.

FACT 377 ☞ There are more than seventy **potentially active volcanoes** in Alaska.

FACT 378 ☞ Alaska has almost twice as many **caribou** as people.

FACT 379 ☞ In 2010, Alaska had the **fifth-highest rate of violent crime** in the United States.

FACT 380 ☞ The most violent volcanic event of the last century was the **1912 eruption of Alaska's Novarupta Volcano**, which created the Valley of Ten Thousand Smokes in Katmai National Park. The other Valley of Ten Thousand Smokes was behind my high-school cafeteria.

FACT 381 ☞ **Three of the ten strongest earthquakes** ever recorded in the world occurred in Alaska.

FACT 382 ☞ Alaska has about five thousand earthquakes **every year**.

FACT 383 ☞ **Suicide** was the sixth leading cause of death in Alaska in 2009.

FACT 384 ☞ Since 1976, Alaska has ranked in the top five states in the nation for the **highest rate of reported rape per capita**.

FACT 385 ☞ Some towns in Alaska **cannot be reached by car**. One is Juneau, Alaska's state capital, which can only be reached by plane, boat or a really big catapult.

FACT 386 Some towns in northern Alaska live in **complete darkness from mid-November to late January**. In the summer, those same towns have no darkness at all from early May to early August.

FACT 387 Alaska's **record temperatures** range from a high of 100°F (Fort Yukon, 1915) to a low of -80°F (Prospect Creek Camp, 1971).

FACT 388 In the winter of 1952–53, Thompson Pass near Valdez, Alaska, got **975 inches of snow**.

FACT 389 In 1989, the **Exxon oil tanker *Valdez*** hit a reef in Alaska's Prince William Sound, spilling more than 11 million gallons (41,600 kilograms) of crude oil over 1,100 miles (1,600 kilometers) of coastline.

FACT 390 There are an estimated 1,000 to 1,200 grizzly bears currently living in the lower forty-eight states; **Alaska has an estimated 30,000 grizzlies.** These will continue to be estimates until they find someone brave enough to go into the woods and count the bears in person.

FACT 391 In the last decade, Alaska has consistently placed among the ten worst states for **male-on-female homicides.**

FACT 392 From 2002 to 2004, Alaska had the nation's highest rate per capita of **men murdering women**, and the state had the second-highest rate per capita from 2005 to 2007.

FACT 393 In 2009, Alaska had **more occupational fatalities** than any other state in America, and the state had the second-highest rate of occupational fatalities in 2010.

FACT 394 The Great Alaskan Earthquake of 1964 measured **9.2 on the Richter scale** and lasted between three and five minutes.

FACT 395 The 1964 Alaskan earthquake was the **strongest in North American history**, and the second strongest ever recorded in the world.

FACT 396 The 1964 quake **generated a tsunami that devastated many towns** along the Gulf of Alaska and caused serious damage along the western coasts of Canada and the United States.

FACT 397 The quake's **death toll was 128**—most of them from the tsunami that followed the actual quake—and it caused $311 million in property loss.

FACT 398 ☞ Waves of up to **sixty-seven meters** were recorded during the tsunamis that followed the 1964 earthquake.

FACT 399 ☞ **Effects of the quake** were felt as far away as the Gulf Coasts of Louisiana and Texas, where seiche action in waterways caused minor damage, and in tide gauges in Cuba and Puerto Rico.

FACT 400 ☞ **Catacombs beneath Paris, France, contain the remains of an estimated 6 million people whose bodies were moved there in the late eighteenth century after the city became overrun with corpses.**

FACT 401 ☞ **Stacked bones and skulls** are used as decorative pieces on walls and doorways in the catacombs.

FACT 402 ☞ Stephen King was inspired to write his classic novel *The Shining* while staying at the **Stanley Hotel in Estes Park, Colorado**, which is said to be haunted.

FACT 403 ☞ Beneath Edinburgh, Scotland, lies a maze of vaults dating back to the mid-1700s. Once a haven for criminal activity, the caverns are now said to be **haunted by ghosts who attack visitors**.

FACT 404 ☞ If you visit New Orleans, Louisiana, you might want to avoid St. Louis Cemetery #1, especially at night. **The graveyard is said to be haunted** by the specter of nineteenth-century voodoo priestess Marie Laveau, who often appears as a black cat with red eyes.

FACT 405 ☞ A pocket of magma deep below Lake Nyos in Cameroon leaks carbon dioxide gas continuously into the water above. **A 1986 explosion on the lake from built-up pressure** created a CO2 cloud that asphyxiated 1,700 people. Experts say such an event could happen again.

FACT 406 ☞ One sustainability company has ranked Miami, Florida, as the **most risky city for natural disasters in the United States**.

FACT 407 ☞ The U.S. Geological Survey estimates the southern tip of Florida can expect **more than sixty hurricanes** over a hundred-year period.

FACT 408 ☞ From 1972 to 1984, **droughts in the Sahel region of Africa** caused more than one hundred thousand deaths.

FACT 409 ☞ In 2005, the rains of Hurricane Stan flooded Central America and southern Mexico, causing more than nine hundred mudslides. **Entire villages were buried**; one village in Panabaj, Guatemala, was declared a cemetery after officials gave up hope of excavating the bodies of three hundred missing villagers.

START SPREADIN' THE "EWS": THE DIRTY TRUTH ABOUT THE BIG APPLE

If you can survive here, you can survive anywhere.

FACT 410 ☛ **New York City has more residents** than the combined populations of Alaska, Vermont, Wyoming, South Dakota, New Hampshire, Nevada, Idaho, Utah, Hawaii, Delaware and New Mexico.

FACT 411 ☛ New York City has **more Irish residents than Dublin**, Ireland; more Italians than Rome, Italy; and more Jews than Tel Aviv, Israel.

FACT 412 ☛ The city is home to more than **a million stray dogs and half a million stray cats.**

FACT 413 ☛ An estimated ten thousand abandoned, orphaned and runaway children **roamed the streets of New York City** in 1852.

FACT 414 ☞ About 1,600 people are **bitten by other people** every year in New York City.

FACT 415 ☞ More New Yorkers **die by suicide** than by homicide.

FACT 416 ☞ About **40 million pounds of dog excrement** were deposited on the streets of New York City every year until a 1978 law made it mandatory for owners to clean up after their pets. Now there are only 30 million pounds a year on the streets.

FACT 417 ☞ It is **illegal to jump off the Empire State Building**. There's a cop posted on the fiftieth floor who hands out citations to people as they fly by.

FACT 418 ☞ It is against the laws of New York City for a man **to leer at a woman**. I know for a fact that this law is not being enforced...

FACT 419 ☞ **Women may go topless in public** in New York City as long as they are not charging money for it.

FACT 420 ☞ New Yorkers cannot dissolve a marriage for irreconcilable differences **unless they both agree to it**. But if they both agree to it, doesn't that mean their differences are reconcilable?

FACT 421 ☞ Because of an ongoing battle with graffiti, New York City made it **illegal to carry an open can of spray paint**.

FACT 422 ☞ The word "hooker" is thought to have originated at **Manhattan's Corlaer's Hook**, a notorious prostitution hot spot in the early nineteenth century.

FACT 423 ☞ More than a third (36 percent) of the current population of New York City was born **outside the United States**.

FACT 424 New York City's **Washington Square Park** was once used as a cemetery and as a site for public hangings.

FACT 425 **The infamous Hangman's Elm still stands** in the northwest corner of Washington Square Park. Look closely and you can see where someone carved, "Archibald was h—"

FACT 426 New York City **remained under British control** even after the Declaration of Independence was signed.

FACT 427 In the 1700s, **New Yorkers owned more slaves** than the citizens of any other city in the British colonies except Charleston, South Carolina.

FACT 428 Roughly **37 million people** visit New York City's Times Square every year. All at the same time.

FACT 429 The first Marxist organization in the Western Hemisphere was **the Communist Club of New York**, formed in 1847 by Joseph Wedemeyer, a friend of Karl Marx and Friedrich Engels.

FACT 430 At **only nine feet wide**, the house at 75 Bedford Street in Greenwich Village is New York's narrowest, and was once the home of poet Edna St. Vincent Millay.

FACT 431 In the late nineteenth century, **thousands of pigs roamed Manhattan** to consume garbage; this was the city's first sanitation system.

FACT 432 America's **first automobile accident** occurred in New York City in 1897.

FACT 433 America's **first auto accident fatality** also occurred in New York City, in 1900.

FACT 434 The top of the Empire State Building was meant to be a **docking port for airships**, but high winds made it too dangerous. Only one airship ever docked there.

FACT 435 In summertime, the surface temperature of an average New York City street can reach **up to 150°F**. Eggs cook at 145°F.

FACT 436 The Empire State Building is estimated to be struck by lightning **about twenty-five times a year**, though some put the figure closer to a hundred.

FACT 437 On an average weekday, more than **5.1 million people ride a New York City subway**. About 3.5 million of them pee in it.

FACT 438 ☞ New York City is estimated to have 8 million rats, **or one rat per person**; others put the figure as high as twelve rats per person.

FACT 439 ☞ The name "hot dog" originated on Coney Island in the late 1800s when a German immigrant named Charles Feltman began selling frankfurters from a cart. His competitors started a rumor that Feltman's frankfurters **contained dog meat**.

FACT 440 ☞ Brazil is plagued with rampant street crime in larger cities like Rio de Janeiro and Sao Paolo. **Tourists are frequent targets.**

FACT 441 ☞ **Common in Brazil are "quicknappings," where a victim is abducted and forced to remove money from an ATM for his own ransom.**

FACT 442 ☞ Colombia is the **world capital for kidnappings**. The government's National Fund for the Defense of Personal Liberty (*Fondelibertad*) estimated that 282 people were kidnapped during 2010 alone.

FACT 443 ☞ The murder rate in Colombia is also one of the world's highest, with **more than 15,200 killings in 2010**. Many victims were local government officials who challenged the drug cartels.

FACT 444 ☞ **Colombia supplies 75 percent of the world's cocaine.**

FACT 445 ☞ **Five Catholic missionaries** were murdered in Colombia in 2005.

FACT 446 ☞ **Kidnappings of foreigners** are prevalent in Russia, as they fetch higher ransoms.

FACT 447 ☞ The presence of Al-Qaeda and the Taliban **makes Pakistan potentially dangerous to Americans** in particular. In 2008, at least sixty suicide bombings killed more than one thousand people.

FACT 448 ☞ The U.S. State Department **warns against all travel to Sudan**. In recent years, Americans and Europeans in Sudan have been victims of robberies and carjackings; two American Embassy employees were assassinated there in 2008.

FACT 449 Nestled precariously between Israel and Syria, **Lebanon is on the U.S. State Department's travel advisory list** due to the presence of Al-Qaeda and Hezbollah militants.

FACT 450 The security threat level in Yemen is high. In 2008, a convoy of tourists was attacked by suspected Al-Qaeda terrorists; two were killed. That same year an **assault on the U.S. Embassy** in Sana'a left an embassy guard dead.

FACT 451 The State Department **restricts the movement of embassy personnel in Algeria** because of suicide car-bomb attacks, kidnappings and assassinations aimed at foreigners.

FACT 452 Somalia's piracy problem and political vacuum—it hasn't had a proper functioning government in about fifteen years—are two reasons why it remains **one of the most dangerous countries in the world.**

FACT 453 Somalia's infamous pirates have been sailing farther afield, into waters where they are less likely to be caught. **"Piracy will only continue, if not increase,"** says one expert. "The business is lucrative and there's no real effective military response."

FACT 454 In 1976, a 7.5-magnitude earthquake **killed twenty-three thousand Guatemalans**.

FACT 455 ☞ More than three thousand people are believed to have perished within the walls of **Beechworth Lunatic Asylum in Victoria, Australia**, victims of abuse, neglect and inhumane medical treatments and experiments.

FACT 456 ☞ One particularly heinous treatment at Beechworth Asylum was the Darwin Chair, a primitive shock therapy device that **spun subjects around so fast that they would bleed from their mouths, eyes, noses and ears**.

FACT 457 ☞ Opened in 1867 and closed in 1995, Beechworth is now the site of **frequent ghost sightings** and offers tours to the public.

FACT 458 ☞ South Africa has been called **the "rape capital of the world"** and also consistently has one of the world's highest homicide rates.

FACT 459 ☞ Ten million South Africans are **infected with HIV**.

DON'T MESS WITH TEXAS. IT'S MESSED UP ENOUGH ALREADY.

Davy Crockett famously said, "Y'all can go to Hell. I'm going to Texas." Did they all end up in the same place?

FACT 460 ☞ The **worst natural disaster in U. S. history** was the Galveston Hurricane in 1900, which killed more than eight thousand people.

FACT 461 ☞ Galveston Island in Texas is reportedly **haunted by the spirits** of the more than eight thousand victims of the Galveston Hurricane.

FACT 462 ☞ Ghosts on Galveston Island are reportedly most often encountered at **Ashton Villa**—a mansion that survived the storm—and **Hotel Galvez**, the island's oldest hotel.

FACT 463 In a 2010 University of Texas/*Texas Tribune* poll, nearly a third of Texans surveyed believed that **humans and dinosaurs roamed the Earth at the same time.**

FACT 464 **More than half of respondents** said they disagree with the theory that humans developed from earlier species of animals.

FACT 465 Almost 40 percent of respondents in the University of Texas/*Texas Tribune* survey agreed with the statement **"God created human beings pretty much in their present form about ten thousand years ago."** Are you picturing a caveman with a cowboy hat and boots and his cavewife with giant bouffant hair and big fake boobs? I am.

FACT 466 Texas passed a law in 2009 that **requires state public schools to incorporate Bible literacy into the curriculum.**

FACT 467 ☞ Almost a third (31 percent) of Texas voters believe that **their state has the right to secede** from the United States and form an independent country.

FACT 468 ☞ Seventy percent of Texans support **offshore oil drilling**; 66 percent support deep-water drilling.

FACT 469 ☞ In Texas, 69 percent of likely voters believe that **the military should be used at the Mexican border** to prevent illegal immigration.

FACT 470 ☞ More than half of Texans (56 percent) support a state lawsuit against the federal government to stop the **recently passed national health care plan**.

FACT 471 ☞ About 70 percent of Texans **support the death penalty**; 49 percent strongly support it. Half of those want to be the ones to flip the switch on the electric chair.

FACT 472 ☞ A 2003 poll found that although a majority of Texans support the death penalty, 69 percent of respondents believe the state has **executed innocent people**.

FACT 473 ☞ **Is the death penalty applied fairly?** In a 2002 *Houston Chronicle* poll, a majority of Texans (59 percent) said yes, although a lower percentage (43 percent) of Americans agrees with them.

FACT 474 ☞ Seventy percent of Texans believe that homosexuality is morally wrong. A majority support amending the U.S. Constitution to **ban gay marriage**.

FACT 475 ☞ **A majority of Texans** (63 percent) support their state's prohibition on recognizing gay marriage.

FACT 476 A 2011 *Texas Tribune* poll showed that while most Texans want to balance the budget by cutting spending, **none of the proposed spending cuts** currently under consideration by the legislature was supported by more than 40 percent of respondents.

FACT 477 The Dallas/Fort Worth airport is so large that it **has its own city designation** (DFW Airport), zip code and public services.

FACT 478 DFW Airport is **larger than Manhattan island**.

FACT 479 In Texas, it is still **a "hanging offense"** to steal cattle.

FACT 480 **In Texas, you can become legally married** by introducing a person as your husband or wife three times in public.

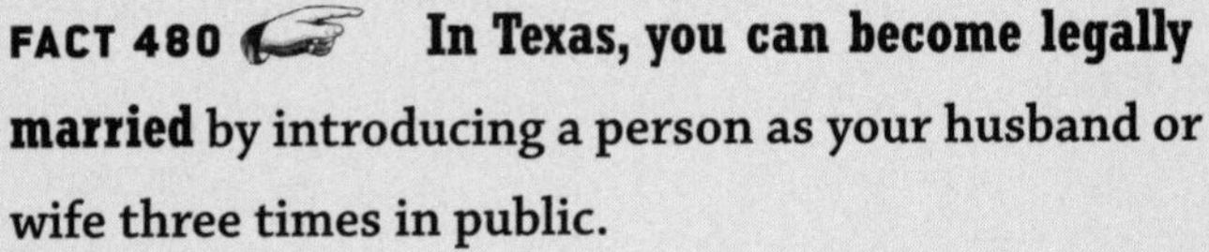

FACT 481 The campus of the University of Texas at Brownsville sits on the site of historic Fort Brown and **is said to be haunted** by ghosts of victims of a nineteenth-century yellow fever outbreak who were buried there.

FACT 482 In the late 1800s, the town of Aurora, Texas, claimed to be **the crash site of an alien spaceship**.

FACT 483 In 1991, Channelview, Texas, mom Wanda Holloway tried to hire someone to kill the mother of a girl who was **competing with Holloway's daughter for a spot on their junior high school cheerleading squad**. Holloway hoped that the girl's grief would make her drop out of the competition.

FACT 484 Holloway became known as **the "Texas-Cheerleader-Murdering-Mom"**; her story gained national attention and spawned two TV movies.

FACT 485 Holloway was convicted of solicitation of capital murder in 1991 and sentenced to fifteen years in prison, but **only served six months** before being released.

FACT 486 There is a town in Texas named **Ding Dong**.

FACT 487 The city of Houston is **slowly sinking** due to the removal of swamp water from deep beneath its surface.

FACT 488 Sam Houston **did not want Austin to be the state's capital**; he called it "the most unfortunate site on earth for a seat of government."

FACT 489 ☞ The city of Henderson has an **outhouse with its own historical marker**. I wonder if it reads, "George Washington shat here"?

FACT 490 ☞ Texas was the **deadliest state for workers in 2010**, with 456 on-the-job fatalities. New Hampshire, with only five workplace deaths, was the safest.

FACT 491 At risk for droughts, floods, earthquakes, landslides, volcanoes and tsunamis, the Indonesian islands of Java and Sumatra face perhaps **more natural disaster hazards than anywhere else in the world**.

FACT 492 The 2004 Indian Ocean tsunami killed more people in Indonesia than in any other country. Of the estimated 228,000 total dead, **57 percent (130,000) were Indonesians.**

FACT 493 From 1907 to 2004 (before the tsunami), Indonesia lost **9,329 people to drought; 17,945 to volcanoes; and 21,856 to earthquakes**.

FACT 494 Because it lies near the North Anatolian Fault, the city of Istanbul, Turkey, and its 12.8 million residents **face high odds of being hit by a major earthquake** in the next twenty-five years.

FACT 495 The region's last big quake was a 7.6-magnitude temblor in 1999 that devastated the city of Izmit. **The official death toll was around seventeen thousand**, but one researcher put the number at forty-five thousand.

FACT 496 A January 2010 study in the journal *Nature Geosciences* found that tensions along the North Anatolian Fault are building and **could trigger numerous small-to-moderate quakes** in the coming years.

FACT 497 A popular destination for scuba divers, **Truk Lagoon in Micronesia is not for the easily spooked**. More than sixty Japanese ships were sunk there by Allied forces in 1944, many with crews trapped inside. Divers of the ruins can spot gas masks, sake cups and the odd "human remain."

FACT 498 ☞ The Winchester "Mystery" House in San Jose, California, built by heirs of the famous gun company, is said to be **haunted by spirits of victims of Winchester rifles** and by the ghost of mad Sarah Winchester herself, who died in the mansion in 1922.

FACT 499 ☞ **The odd, labyrinthine construction** of the 160-room Winchester House adds to its creepiness: staircases lead into the ceiling, doors open to blank walls, spider motifs abound, and candelabras, coat hooks and steps are arranged in multiples of thirteen.

FACT 500 ☞ Once a thriving city built around the Chernobyl Nuclear Power Plant in Ukraine, Pripyat was evacuated during the 1986 disaster and is now an eerie ghost town with **still-dangerous levels of radiation**.

FACT 501 ☞ Originally planned as a vacation resort catering to American servicemen, the Sanzhi District in New Taipei, Taiwan, was abruptly abandoned in 1980 after numerous **freak construction accidents were attributed to supernatural causes**.

FACT 502 In 1944, the French town of Oradour-sur-Glane was destroyed by Nazis, who also **slaughtered all 642 residents.** Today the abandoned, half-burned village stands as a memorial to the tragedy.

FACT 503 At the Chapel of Bones in Portugal, **walls and columns are covered in artistic designs of bones** from more than five thousand exhumed skeletons. On one wall a child's dried corpse hangs from a chain.

FACT 504 In 2011, St. Louis, Missouri, **topped the FBI's list of the most dangerous cities in America**, based on statistics for murder, rape, robbery and assault.

FACT 505 St. Louis **beat out perennial favorites Detroit and New Orleans** to take the dubious prize.

FACT 506 **Kidnapping, murders, death threats, drug-related shoot-outs, carjacking, armed robberies and home break-ins are common in Port-au-Prince, Haiti.**

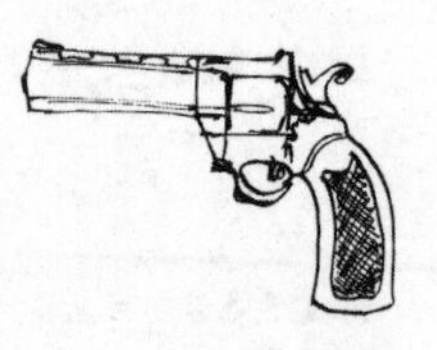

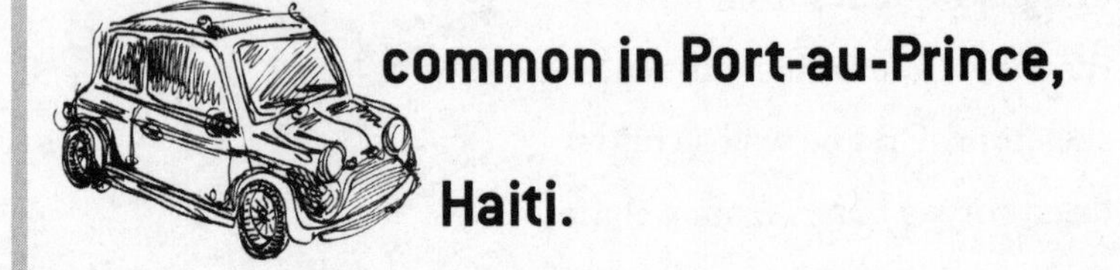

FACT 507 The murder rate in Caracas, Venezuela, is said to be among the highest in the world, with much of the violence **related to drug trafficking**.

FACT 508 ☞ Port Moresby in Papua New Guinea is the home of **frequent rapes, armed robberies and carjackings**. Visiting unguarded public sites such as golf courses, beaches, parks or cemeteries can be dangerous for visitors.

FACT 509 ☞ While mugging and pickpocketing are the most common crimes against tourists in **Santo Domingo, Dominican Republic**, reports of violence against both locals and foreigners are growing.

FACT 510 ☞ A corrupt government and the huge gap between the rich and the poor in Guatemala City, Guatemala, make it **one of the most dangerous cities in the world**.

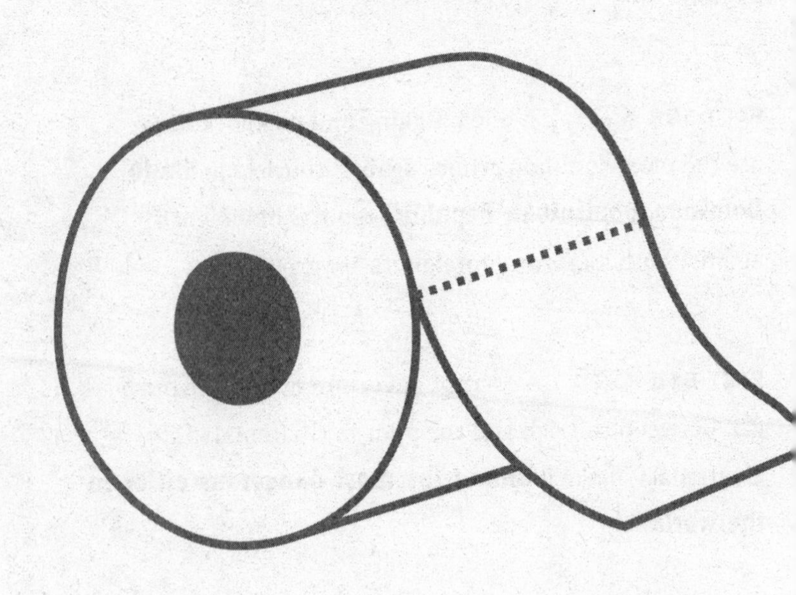

THE BETTER TO EAT YOU WITH, MY DEAR

Dangerous Animals Make Lousy Pets

I LOVE IT WHEN people talk about how humans sit at the top of the food chain. You often hear this proclamation when the subject of vegetarianism or PETA comes up and people feel the need to defend their flesh-eating ways.

"Why shouldn't we eat meat?" they will say. "After all, we are at the top of the food chain." Right—we conquered the mighty chicken! How hard is that? They can't even fly. And cows. Ever seen a cow run? Neither have I.

Go over to Africa and hang out with the lions and hippos for a few weeks, or with the grizzlies up in Alaska, or the great white sharks off the coast of South Africa. Then come back to me and let's talk about the food chain. Oh, wait, you can't—you'll be dead.

FACT 511 Nile crocodiles, the largest in Africa, **can grow up to 20 feet** and 1,650 pounds.

FACT 512 Nile crocs are estimated to **kill as many as two hundred people** every year.

FACT 513 After infant Azaria Chamberlain was allegedly taken from her tent by a dingo near Ayers Rock, Australia, in 1980, her mother, Lindy, was convicted of the child's murder. **Chamberlain served four years in prison before her conviction was overturned** after a piece of Azaria's clothing was found in a dingo den near the site of her disappearance.

FACT 514 **In 2001, a nine-year-old boy was attacked and killed by two dingoes** on Fraser Island, Australia.

FACT 515 ☞ While piranhas aren't quite the vicious man-eaters of myth, **attacks on humans by these fierce fish are increasing in frequency,** often resulting in lost fingers or toes.

FACT 516 ☞ In September 2011, **more than one hundred swimmers at a popular lake in Brazil** were attacked by piranhas, many requiring hospital treatment for their injuries.

FACT 517 ☞ When an aquarium in Wales went to great lengths to acquire a male and a female piranha (piranhas are illegal to import in most parts of the world, including Britain) in hopes that the two would mate, **the female ate her potential suitor**. I can't decide if that's the ultimate compliment or the ultimate rejection.

FACT 518 ☞ The largest brown bear species, the Kodiak, can weigh up to 1,700 pounds and rivals the polar bear as **the world's largest land-based predator**.

FACT 519 ☞ **A type of brown bear**, the grizzly is an omnivore that can grow up to eight feet tall and eight hundred pounds.

FACT 520 ☞ Brown bear attacks often result in serious injury and, in some cases, **death**.

FACT 521 ☞ Bears kill an average of two people per year in North America. The majority of fatalities result from **humans coming between a mother and her cubs**.

FACT 522 ☞ A brown bear's physical strength is so great that **a single bite or swipe can be deadly**. Some human victims' heads have been completely crushed by a bear bite.

FACT 523 ☞ In July 2011, a tourist at Yellowstone Park was killed when he and his wife **surprised a mother grizzly bear while hiking**. The bear charged as they tried to leave the area, killing the man. His wife played dead and escaped without serious injury.

FACT 524 ☞ In 2003, bear activist Timothy Treadwell and his girlfriend Amie Huguenard were **killed and eaten by grizzly bears in Katmai National Park, Alaska**. The two had spent several months in the bush and were scheduled to leave the next day.

FACT 525 ☞ **The great white shark is the largest predatory fish on earth,** averaging fifteen feet in length, though specimens exceeding twenty feet and weighing five thousand pounds have been recorded.

FACT 526 ☞ As many as half of the one-hundred-plus annual shark attacks worldwide are **attributable to great whites**.

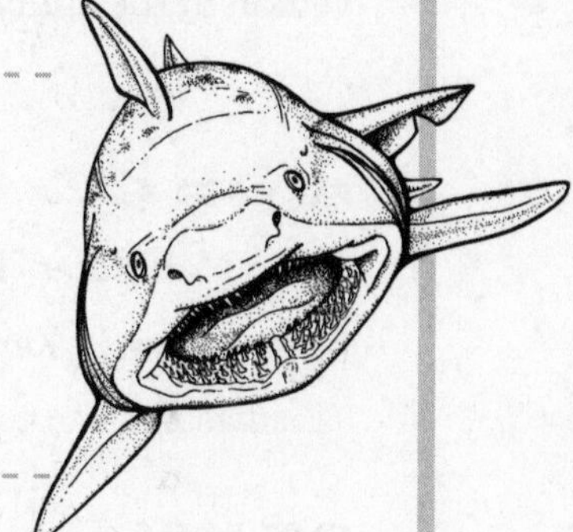

FACT 527 ☞ The United States has **more reported shark attacks than any other country**, with a total of 1,049 attacks—49 fatal—from 1670 to 2009.

FACT 528 ☞ As of 2009, the International Shark Attack File had recorded a total of 2,251 attacks worldwide since 1580, with **more than 20 percent of them resulting in fatalities**.

FACT 529 ☞ More recorded shark attacks have occurred in **New Smyrna Beach, Florida**, than anywhere else, earning it the title of "shark bite capital of the world." I wonder if they put that on a bumper sticker.

FACT 530 ☞ When **emus began destroying thousands of acres of wheat crops** in Western Australia after World War I, the government ordered the military to use machine guns to cull the hungry birds' massive numbers.

FACT 531 The box jellyfish is equipped with about **five thousand stinging cells** that pack the most deadly venom in the animal kingdom.

FACT 532 In 1898, **a pair of lions** known as the Tsavo Man-Eaters **killed a reported 135 people** working on a railroad in Kenya.

FACT 533 The venomous bite of **the brown recluse spider** can lead to necrosis (skin tissue death).

FACT 534 **Though vampire bats feed primarily on the blood of cattle and horses, they will attack humans.**

FACT 535 The Venus flytrap catches its prey in a tenth of a second but **takes ten days to fully digest** its unfortunate meal.

FACT 536 The **largest manta ray** ever recorded was more than twenty-five feet wide.

FACT 537 When faced with a predator, **the 1,500-pound cape buffalo will charge head-on**, leading with its two big, sharp horns.

FACT 538 A master of concealment, the stone fish lives in coastal areas of the Indian and Pacific Oceans. Step on it and you can get a sting from one or more of its **venomous spines, which cause excruciating pain** and, in some cases, death.

FACT 539 ☞ The Inland Taipan in the **most venomous land snake in the world**, injecting its victims with poison fifty times more toxic than that of a cobra.

FACT 540 ☞ The Portuguese Man-of-War is not a jellyfish but a siphonophore, a colony of organisms working together. With **tentacles up to 165 feet long**, the Man-of-War's sting can kill a human.

FACT 541 ☞ At a **weight of up to eight thousand pounds**, the hippopotamus is the heaviest land mammal after the elephant, but it can gallop at 18 mph.

FACT 542 ☞ The hippopotamus is one of Africa's most dangerous animals. Extremely aggressive, unpredictable and unafraid of humans, **hippos will upend boats without provocation and bite occupants** with their huge sharp teeth.

FACT 543 ☞ The hippopotamus's canine teeth and incisors grow continuously, with canines reaching **twenty inches (fifty-one centimeters) in length**.

FACT 544 ☛ **Most human deaths by hippopotamus** occur when the victim gets between the animal and deep water or between a mother and her calf.

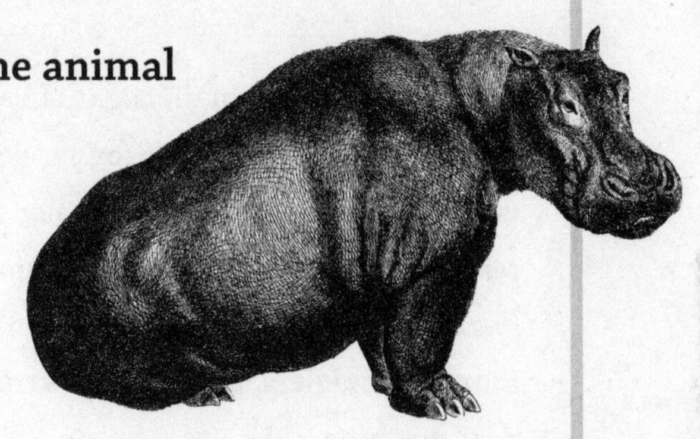

FACT 545 ☛ Many famous African explorers and hunters, such as Livingstone, Stanley, Burton and more, had boating mishaps with hippos and **considered them to be dangerous beasts**.

FACT 546 ☛ In the early 1970s, while hunting near the shores of Lake Rukwa, Tanzania, a hunter was killed when a bull hippo turned over his canoe and **bit off his head and shoulders**.

FACT 547 ☞ Estaurine crocodiles are the most prolific man-eaters on earth, **killing approximately two thousand people a year**.

FACT 548 ☞ On the night of February 19, 1945, crocodiles were responsible for perhaps the most devastating animal attack on humans in recorded history, reportedly feasting on retreating Japanese troops in a swampy area off the coast of modern-day Burma. **Of one thousand soldiers, only twenty were found alive the next morning**.

FACT 549 ☞ With **ten arms and a body that can reach up to sixty feet in length**, the giant squid is a carnivorous predator. One notable incident occurred on March 25, 1941, when a survivor of the sunken British ship *Britannia* was pulled under water and presumably killed by a giant squid.

FACT 550 ☞ A male squid will sometimes **eat the female after mating**.

FACT 551 ☞ **Squids can fly**: they are able to propel themselves out of the water in the same way they swim, by forcing water through their bodies at high velocity.

FACT 552 ☞ Squid can propel themselves in flight to **distances of up to 164 feet**.

FACT 553 ☞ The world's largest lizard, the Komodo dragon can reach ten feet in length and weigh more than three hundred pounds. **It is the top predator** on the Indonesian islands where it lives.

FACT 554 ☞ The Komodo dragon's diet normally consists of deer, wild goats and pigs, but **the animal will eat anything it can catch, including humans**.

FACT 555 ☞ Komodo dragons **devour their prey completely, including the bones**. After a French tourist was killed in 1986 by a Komodo, the only evidence left of him was his bloodstained shoes.

EL GATO LOCO: FREAKY CAT FACTS

Example: They spend 85 percent of their waking hours silently plotting your demise.

FACT 556 A cat's nose pad is as unique as a human fingerprint. **No two nose prints are identical**. I think I saw this on *CCSI* (*Cat Crime Scene Investigation*).

FACT 557 In the animal kingdom, **a cat's intelligence** is surpassed only by that of monkeys and chimps. If cats ever figure out how to fling their poop, we're in big trouble.

FACT 558 **A cat's hearing is much stronger and more sensitive than a human's or a dog's.** Humans can only hear sounds up to 20 khz, but a cat can hear up to 65 khz.

FACT 559 ☞ A cat's **sense of smell** is fourteen times stronger than a human's.

FACT 560 ☞ Every year, nearly four million cats are **eaten in Asia**.

FACT 561 ☞ The ancient Egyptians **were not the first to domesticate cats**. The oldest known pet cat was recently found in a 9,500-year-old grave in Cyprus, predating Egyptian art depicting cats by 4,000 years or more.

FACT 562 ☞ Cats were **associated with witchcraft in the Middle Ages**. On holy days, Europeans would throw the animals from church towers or into bonfires. Sometimes they did it on other days, too, just for fun.

FACT 563 ☞ The first cat in space was launched by France in 1963. Felicette (aka "Astrocat") had electrodes implanted in her brain to send neurological signals back to Earth. **She survived the trip**.

FACT 564 ☞ Approximately forty thousand Americans are **bitten by cats** every year.

FACT 565 ☞ A cat can **travel at a top speed of approximately 31 mph** (49 km) over a short distance. When spun by the tail and flung.

FACT 566 ☞ Black cats **are considered good luck** in Britain and Australia.

FACT 567 ☞ A cat can jump up to **five times its own height** in a single bound.

FACT 568 ☞ The **largest known litter** of cats ever produced was nineteen kittens.

FACT 569 ☞ **The heaviest cat** on record was a forty-seven-pound tabby (flabby?) from Queensland, Australia, that died in 1986 at age ten.

FACT 570 ☞ A cat typically can live up to 20 years, which is equivalent to about 96 human years. The **oldest cat on record died in 2005 at the age of 38, or 169 in human years**.

FACT 571 ☞ In 1888, more than three hundred thousand mummified cats were found in an Egyptian cemetery. They were **used as fertilizer by farmers** in England and the United States.

FACT 572 ☞ How do lost cats find their way home? Experts think they either use the angle of the sunlight or **have magnetized cells in their brains** that act as compasses.

FACT 573 ☛ The world's rarest coffee, Kopi Luwak, is **made from the dung of Indonesian wildcats** and sells for $500 a pound. Guess what it tastes like?

FACT 574 ☛ In just seven years, a single pair of cats and their offspring could produce a **whopping 420,000 kittens**.

FACT 575 ☛ Cats **spend nearly a third of their waking hours** cleaning themselves.

FACT 576 ☛ The earliest ancestor of the modern cat lived about **30 million years ago**.

FACT 577 ☛ Cats are **extremely sensitive to vibrations**; some believe they can detect earthquake tremors ten or fifteen minutes before humans. But do they warn us? Of course not. They're cats, and cats are assholes.

FACT 578 ☞ There are more than 500 million domestic cats in the world, with around forty recognized breeds.

FACT 579 ☞ There are as many as **60 million feral cats** in the United States alone.

FACT 580 In the central provinces of India, **leopards will enter native huts to find their prey.**

One, known as the Panawar Man-eater, is believed to have killed four hundred people.

FACT 581 In October 1943, a lone lion was shot in what is now Zambia **after it had killed forty people**.

FACT 582 Pumas have been known to catch prey **seven to eight times their size**.

FACT 583 ☞ Since 1970 there have been **more than forty puma attacks on humans**; at least seven of them were fatal.

FACT 584 ☞ In 1994, **two female joggers in California** were killed and eaten by pumas.

FACT 585 ☞ In 1992, a group of children playing in a mango plantation in South Africa were attacked by a **twenty-foot python, which swallowed one of them whole**.

FACT 586 ☞ A tigress known as the Champawat Man-eater **killed 438 people in Nepal** between 1903 and 1911.

FACT 587 ☞ **A relative of the squid and the octopus, the cuttlefish has ten arms to catch prey and a beak to inject its catch with a fast-acting venom.**

FACT 588 The hooded pitohui is a South Pacific songbird **whose skin and feathers contain the same poison found on South American dart frogs**. The poison can be transferred to humans who handle the birds.

FACT 589 Abundant in the United States, the centipede carries venom that is not fatal to humans, **though some large species pose a danger to children.**

FACT 590 **Some ocean coral are poisonous**; the most deadly of these is the Palythoa. Its sting can cause chest pains, breathing difficulty, racing pulse and death within minutes. There is no treatment.

FACT 591 Two species of giant catfish **are reported to be man-eaters**: the wels in Europe and the goonch in Asia.

FACT 592 ☞ Scientists believe male bottlenose dolphins, which live in shallow waters around the world, **can be incited to attack young girls** because of the strong hormones they release.

FACT 593 ☞ The slow loris is an adorable but deadly primate that **emits toxins from its elbows**, making it one of the only poisonous mammals in the world.

FACT 594 ☞ The slow loris takes toxin in its mouth just before it bites. In humans, the bite can cause **anaphylactic shock and death**.

FACT 595 ☞ Giant anteaters will attack when threatened, and have extremely sharp four-inch claws that can **kill a human with one swipe**.

FACT 596 When its territory is threatened, **the giant anteater can fight off the fiercest predators,** including pumas and jaguars.

FACT 597 With exceptionally strong jaws and long teeth, **leopard seals are top predators in the Antarctic food chain** and have been reported to hunt humans.

FACT 598 The stingray's barbed tail delivers mild venom that typically causes little more than pain in its victims, but **a sting can kill if the wound becomes infected** or if the barb stabs a vital organ.

FACT 599 TV's **Steve "Crocodile Hunter" Irwin died in 2006** after a stingray's spine pierced his heart.

HONEY BADGER DON'T CARE

He just takes what he wants.

FACT 600 ☛ The ratel, or honey badger, earns its name by attacking beehives to feast on honey and larva, **even as it is stung repeatedly**.

FACT 601 ☛ The South Africans have a saying: **"*so taai soos a ratel,*"** which means, "as tough as a honey badger."

FACT 602 ☛ Known for its fearlessness, **the ratel will attack and eat almost anything**, including animals many times its size, like buffalo, gnu and waterbuck.

FACT 603 The honey badger will also **attack and eat poisonous snakes and insects**. Its thick, loose hide protects it from bites and stings.

FACT 604 Honey badgers are aggressive when threatened, and have been known to **attack humans and even cars**. Nobody said they were smart, just fearless.

FACT 605 Honey badgers, especially wounded ones, **will spray a foul-smelling discharge** from their anal glands to repel enemies.

FACT 606 Because of its ferocity and thick, tough skin, the ratel has **few natural predators**.

FACT 607 The honey badger has a reputation for killing animals **by attacking the scrotum** until the victim bleeds to death.

FACT 608 The honey badger has powerful jaws and sharp teeth that **can bite through animal hide, crocodile skin and turtle shells.**

FACT 609 The honey badger's **skin is loose enough** that the animal can twist itself around to bite an attacker even when it is being held by the back of the neck.

FACT 610 The ratel is a cousin of the equally fierce **American wolverine.**

FACT 611 Snakes make high-yield meals, and **ratels track them relentlessly**. Wherever snakes try to hide—in trees, in dense brush or underground—badgers will follow and attack. This includes lethal species like the cobra and puff adder.

FACT 612 In summer, **snakes provide more than half the food a honey badger consumes.**

FACT 613 ☞ Ratels have been observed **taking kills away from lions.**

FACT 614 ☞ In populated areas, **honey badgers have been known to attack domestic animals** such as sheep and goats, and can become a problem for farmers and ranchers.

FACT 615 ☞ A **wild honey badger can live up to twenty-four years**; one animal in captivity lived over thirty-one years.

FACT 616 ☞ An average male ratel can trot at six miles an hour and patrol **a home range of two hundred square miles (518 square kilometers) or more.**

FACT 617 ☞ To protect her cub from predators, a honey badger mother will move to a new den **every three to five days**, each a mile or more from the last.

FACT 618 ☞ Honey badgers find most of their prey underground, thanks to long claws that grow throughout their lives and enable them to **dig exceptionally well**.

FACT 619 ☞ Even if a ratel is bitten by a venomous snake on its head or another vulnerable spot, it can **sleep off the effects of the poison** with no long-term adverse effects.

FACT 620 ☞ Ratels have been known to **remove the tongue, eyes and brain** of prey.

FACT 621 The pufferfish is the second most poisonous vertebrate in the world and **can kill a human**.

FACT 622 Pufferfish toxin, which has no cure, **paralyzes the diaphragm and causes suffocation**.

FACT 623 The cone snail may look harmless, but this Indo-Pacific native **uses a harpoon-like tooth to inject venom** that paralyzes instantly and can lead to death.

FACT 624 There is **no antivenin** for cone snail bites.

FACT 625 ☞ The geography cone snail (*Conus geographus*) **is the most venomous of five hundred cone snail species**.

FACT 626 ☞ In some parts of the world, **army ants are used as surgical sutures**. The ant seizes the edges of the wound in its powerful jaws before its body is removed, leaving the head and jaws in place to close the wound.

FACT 627 ☞ **Capuchin monkeys have been taught the concept of money**. After researchers taught them to trade silver coins for food, the monkeys rejected higher-priced items for larger quantities of lower-priced snacks.

FACT 628 ☞ The monkeys were even taught to gamble; some **learned to trade coins for sex**.

FACT 629 ☞ A recent study found that **chickens can anticipate future events** and that they demonstrate self-control in the present, a trait previously attributed only to humans and other primates.

FACT 630 The finding suggests that domestic fowl, *Gallus gallus domesticus*, are **intelligent creatures that might worry**.

FACT 631 **Elephants are frightened by pig squeals.** Ancient combatants, like the Romans, used "war" pigs to scare off enemy elephants.

FACT 632 The invasive northern snakehead fish is a **voracious top-level predator with no natural enemies**, which can decimate populations of native fish.

FACT 633 Native to Madagascar, **the aye-aye is an endangered primate** believed by locals to be an omen of bad luck. It is often killed on sight.

FACT 634 Contact with the **venomous puss moth caterpillar** can result in headaches, nausea, vomiting, severe abdominal distress, shock or respiratory stress. Symptoms can last up to five days.

FACT 635 ☞ **The gray wolf is the world's biggest, most powerful dog.** Wolves primarily hunt deer, moose and bison, but when wild supplies are tight, domestic cattle and sheep are easy targets—hence the wolf's uneasy relationship with humans.

FACT 636 ☞ **Despite its whimsical appearance, the male platypus can be dangerous to man. Hidden ankle spurs can deliver a venomous prick that won't kill you, but will cause excruciating pain.**

FACT 637 ☞ **Platypus venom also causes hyperalgesia**, a condition in which the body becomes more sensitive to pain.

FACT 638 Moose are among the most dangerous regularly encountered animals in the world. Though they typically avoid human contact, **a threatened moose will charge aggressively and can cause injury or death**.

FACT 639 **More people are attacked by moose** every year than by bears.

FACT 640 Armed with **powerful jaws, sharp claws and a thick hide**, the wolverine can take down prey as large as a moose and steal food from bears and wolves.

FACT 641 **Elephants are among the most dangerous and aggressive animals in the world, killing hundreds of people each year.**

FACT 642 A threatened rhinoceros can charge at speeds of **up to 35 mph.**

FACT 643 **In 2010, a six-ton killer whale at SeaWorld in Orlando, Florida, grabbed a trainer by the hair and dragged her into the water, killing her.**

FACT 644 The whale that attacked its SeaWorld trainer had been involved in **two other human deaths prior to that incident**.

FACT 645 **Male sea horses, not females**, get pregnant and give birth.

FACT 646 A scent released from **glands in a squirrel's foot** helps it mark its territory.

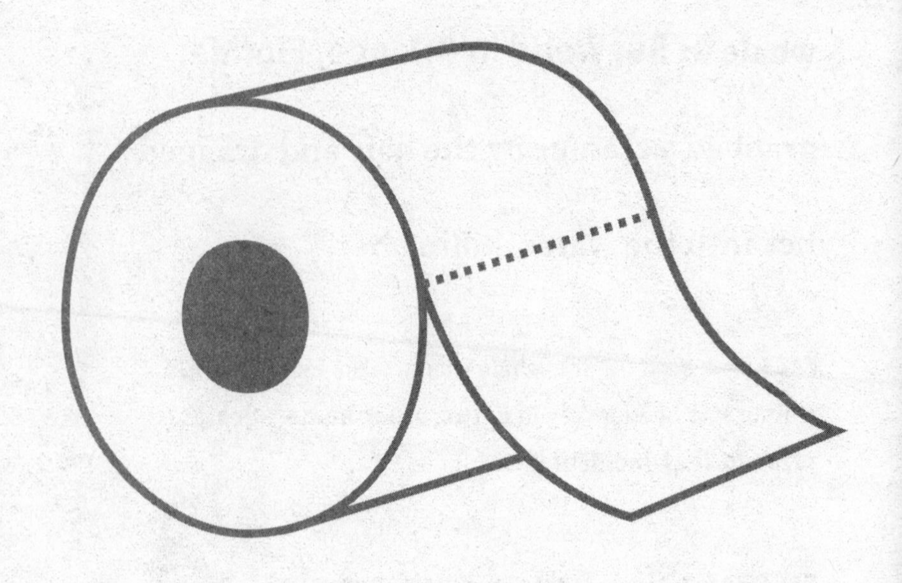

CORPUS HORRIFICUS

Our Bodies, Our Nightmares

THE HUMAN BODY IS a lot like sausage: the more you know about it, the scarier it gets. Take the pancreas, for example. A lovely organ, but you don't want to know what all is going on in there, trust me. Or bacteria—trillions of tiny creatures crawling over every inch of our bodies every minute of every day. And don't even get me started on things like bad hygiene or plastic surgery or disease or what we're doing to ourselves with our bad habits.

If you aren't freaked out by your body, you're not paying attention. It's time to pay attention.

FACT 647 **Antibacterial soap** is no more effective at preventing infection than regular soap, and triclosan (the active ingredient) **can interfere with your sex hormones.**

FACT 648 **The human body is home to some one thousand species of bacteria. There are more germs on your body than people in the United States.**

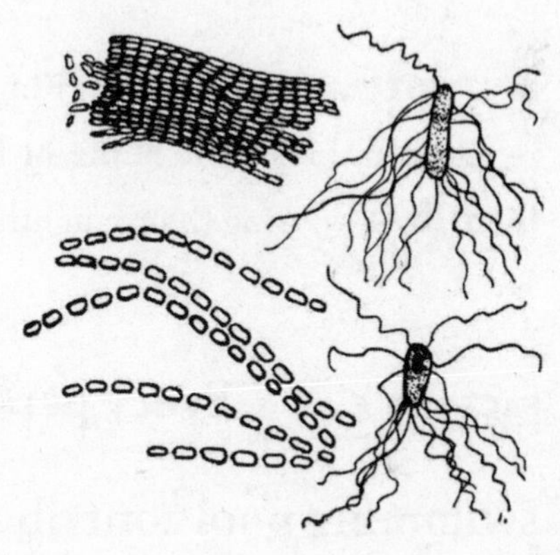

FACT 649 A study of more than eleven thousand children determined that **an overly hygienic environment *increased* the risk of eczema and asthma**.

FACT 650 ☞ Monks of India's Jain Dharma sect are **forbidden to bathe**, as bathing destroys the lives of millions of microorganisms living on the human body.

FACT 651 ☞ Ancient Egyptians **treated cuts and burns with urine**, as urea is known to kill fungi and bacteria.

FACT 652 ☞ England's King Henry IV required his knights to **bathe at least once in their lives**—during their knighthood ceremonies.

FACT 653 ☞ **Every person who enters a swimming pool contributes an average of at least 0.14 grams of fecal material to the water, usually within the first fifteen minutes of entering. And why not? They're already peeing—might as well make it a twofer.**

FACT 654 ☞ Excrement dumped from windows into the streets in eighteenth-century London **contaminated the city's water supply and forced locals to drink gin instead.** Ironic, since gin has always tasted like excrement water to me.

FACT 655 ☞ A Florida seventh-grader recently won her school science fair by proving there are **more bacteria in ice machines at fast-food restaurants than in toilet bowl water**.

FACT 656 ☞ Sorry, but **the "five-second rule" for dropped food is fiction**. Bacteria can contaminate food on contact.

FACT 657 ☞ **Regular teeth brushing** didn't become commonplace in the United States until it was enforced on soldiers during World War II.

FACT 658 ☞ Northern Tissue proudly **introduced "splinter-free" toilet paper in 1935**. Previous wiping options included tundra moss (Eskimos), a sponge with salt water (Romans) and corncobs in the American West.

FACT 659 ☞ NASA recently spent more than $23 million to design a toilet for the space shuttle **that would work in zero gravity**.

FACT 660 ☞ More than a third of Americans **do not see a dentist every year**. The American Dental Association recommends that we all see a dentist twice a year.

FACT 661 ☞ In 2009, almost **half of all Mississippians** did not see a dentist at all.

FACT 662 ☞ In a survey, men, African-Americans, Hispanics and seniors were all **less likely than others to have visited a dentist** in the past year.

DON'T STAND SO CLOSE TO ME: THE HEARTBREAK OF HALITOSIS

What did you have for lunch, a zombie?

FACT 663 ☛ **Halitosis (bad breath) has many causes**, including lack of hygiene, dentures, gum disease, postnasal drip, infection, dry mouth, side effects of medication, smoking, drinking, eating strong foods like garlic and onion, and more.

FACT 664 ☛ Bad breath **can also be caused by diseases** and conditions such as pneumonia, bronchitis, chronic infections, diabetes, acid reflux and liver and kidney problems. But if you've got all that going on, good breath probably isn't a priority.

FACT 665 ☛ **Smoking or chewing tobacco-based products** can cause bad breath, stain your teeth, inhibit your ability to taste foods and irritate your gums.

FACT 666 ☞ **Chocolate causes bad breath**, possibly due its high level of sugar, which can stimulate the growth of odor-causing plaque.

FACT 667 ☞ Foul breath can also be caused by **constipation**, hence the term *dookie breath*.

FACT 668 ☞ Halitosis can also be caused by **bacteria between teeth or under the gums**.

FACT 669 ☞ Xerostomia (dry mouth) can cause bad breath: saliva cleans the mouth by neutralizing acids caused by plaque and washing away **dead cells that accumulate and can decompose** and produce a bad smell.

FACT 670 ☞ Even good oral hygiene—brushing, flossing, mouthwash—cannot get rid of bad breath caused by strong foods like garlic or onions. The **odor will not go away** until the foods have passed through your body.

FACT 671 ☞ Persistent bad breath or a bad taste in your mouth can be **warning signs of periodontal (gum) disease.**

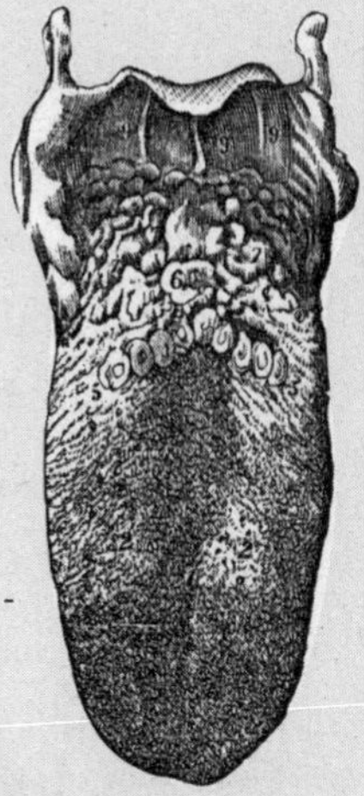

FACT 672 ☞ The tongue is **the most common location in the mouth for bad breath**; bacteria there live on food remnants, dead skin cells and phlegm from postnasal drip, producing smelly sulfur compounds.

FACT 673 ☞ Your tonsils have deep holes where **a foul cheese-like substance can form and cause halitosis.** Some people call this throat feta, but I wouldn't eat it.

FACT 674 ☞ More than 70 percent of teens say **bad breath is an instant turnoff.** Another 85 percent believe that it's the most important thing to avoid when meeting someone for the first time.

FACT 675 ☞ Treating bad breath is big business. **Americans spend nearly $3 billion a year** on gum, mints and mouth rinses. Some Americans need to spend more.

FACT 676 ☞ A 2009 study published in the *Australian Dental Journal* found that "alcohol-containing mouthwashes **contribute to the increased risk of [the] development of oral cancer**." The ethanol in mouthwash is thought to allow cancer-causing substances to permeate the lining of the mouth.

FACT 677 ☞ Researchers from Tel Aviv University have published a study that finds a **direct link between obesity and bad breath**: the more overweight you are, the more likely your breath will smell unpleasant to those around you.

FACT 678 ☞ During the making of *Gone With the Wind*, **Vivien Leigh hated kissing Clark Gable because of his bad breath**. Gable already had a full set of dentures by age thirty-two, which no doubt contributed to his halitosis.

FACT 679 ☞ **Adolf Hitler was terrified of the dentist**; his avoidance of dental care led to abscesses, gum disease and "terribly bad breath." Shame, since he had so many other good qualities.

FACT 680 ☞ In 2009, Marcia Gilmore of Kingston, Jamaica, was fined $6,000 for assaulting a woman who Gilmore said **had chronic halitosis and ignored warnings not to talk to her.**

FACT 681 ☞ During his 1989 appearance on the TV talk show *Larry King Live*, Donald Trump asked King, "**Do you mind if I sit back a little bit because your breath is very bad.** It really is."

FACT 682 ☞ As many as 25 percent of all women giving birth in European and American hospitals from the seventeenth through the nineteenth centuries **died of puerperal fever, an infection spread by unhygienic nurses and doctors**.

FACT 683 ☞ **Television remote controls are the worst carriers of bacteria in hospital rooms; they spread antibiotic-resistant *Staphylococcus*, which contributes to the ninety thousand annual deaths from infection acquired in hospitals.**

FACT 684 ☞ History has long held that President James Garfield died from an assassin's bullet, but researchers now believe that Garfield died from **a severe infection acquired from the manure-stained hands of the medical team** that treated his gunshot wound.

FACT 685 ☞ Researchers in London estimate that **regular hand-washing could prevent a million deaths a year.**

FACT 686 ☞ In a study of sixteen elementary schools and six thousand students, regular use of hand sanitizer in the classroom **decreased absenteeism due to infection** by almost 20 percent.

FACT 687 ☞ More than half of healthy people have *Staphylococcus aureus* **living in their nasal passages, throats, hair or skin**.

FACT 688 The CDC reports that **the spread of pinworms among children** can be reduced by clipping nails and showering children immediately after they wake in the morning. Yeah right. I'd like to see them try that with my kid.

FACT 689 A study by an Australian razor company found that the ladies there **associate facial hair with sexual wildness**.

FACT 690 **Not only is daily bathing unnecessary**, but this relatively recent phenomenon strips your skin of health-promoting bacteria and oils. Do it anyway.

FACT 691 **Would it kill you to floss? It might kill you not to. Studies show that you're twice as likely to have heart disease if you have periodontal disease.**

FACT 692 An estimated **6 to 12 million head lice infestations occur each year** in the United States among children three to eleven years of age.

FACT 693 Yes, **you can get pubic lice ("crabs") from toilet seats,** but it is extremely rare, since lice cannot live long away from a warm human body. Also, their feet are not designed to hold on to smooth surfaces.

FACT 694 British teeth may not be as straight or white as American teeth, **but they are more healthy**. A worldwide dental study found that the average twelve-year-old in Britain has fewer decayed and missing teeth and fewer fillings than his twelve-year-old peers in other Western countries, including America.

FACT 695 ☞ In a study of uncircumcised males, **a third of foreskins contained strains of human papillomavirus (HPV)**, one of the most common sexually transmitted infections, and the cause of 70 percent of cervical cancers.

FACT 696 ☞ HPV can also lead to **genital warts** in both sexes.

FACT 697 ☞ Balanitis, an inflammation of the head of the penis, **occurs most often in uncircumcised men** and boys with poor hygiene.

FACT 698 ☞ Balanitis can also lead to phimosis, a condition in which the foreskin of the penis is so tight that it cannot be retracted. Phimosis makes it **difficult to keep this area clean and can lead to irritation and infection**.

FACT 699 ☞ The **exact cause of penile cancer is unknown**, but the rare disease has several known risk factors, including circumcision, smoking, HPV infection and phimosis.

FACT 700 ☞ Most cases of penile cancer occur in **men over fifty**.

FACT 701 ☞ The most common and effective treatment for penile cancer is surgery, including a **partial or full penectomy (amputation of the penis)**.

FACT 702 ☞ Bacterial vaginosis, **the most common type of vaginal infection**, occurs more often in women who are sexually active, though it is not considered an STD.

FACT 703 ☞ Although its "fishy" smell and vaginal discharge are little more than bothersome for most women, **bacterial vaginosis has been linked to pregnancy complications** and infection after pelvic surgery.

FACT 704 ☞ Risk of bacterial vaginosis **increases if you smoke, douche or have more than one sex partner.**

FACT 705 ☞ **A woman's risk for vulvovaginal candidiasis**, or vaginal yeast infection, increases with several factors, including the use of antibiotics, birth control pills, hormone replacement therapy or corticosteroids; being pregnant; douching; and wearing tight-fitting clothes.

FACT 706 ☞ The ancient Roman historian **Pliny wrote that the "horrible" smell of menstrual blood drives dogs mad**, and even ants will throw away grains of corn that have been touched by it. Why did their corn have menstrual blood on it? Never mind, I don't want to know.

FACT 707 ☞ As recently as the 1920s, **women were encouraged to douche with Lysol** for hygiene and to prevent pregnancy.

FACT 708 ☞ Though the **recent trend of pubic hair removal** appears to have contributed to decreases in the occurrence of pubic lice, cases of gonorrhea and Chlamydia have increased over the same period, a correlation that may not be merely coincidental.

FACT 709 ☞ **Waxing creates inflammation**, which can trap bacteria beneath the skin and promote ingrown hairs and infections like staph and folliculitis.

FACT 710 ☞ You can **contract a sexually transmitted disease during a bikini wax**, from unclean tools or microorganisms that live in the wax and can be passed from visitor to visitor.

FACT 711 ☞ In 2007, an Australian woman with type 1 diabetes **almost died of a bacterial infection** she got from a bikini wax.

FACT 712 ☞ In 2009, **New Jersey lawmakers considered a ban on body waxing** after two women claimed they had contracted infections from the procedure. Then they thought about what a New Jersey beach would look like with unwaxed bodies and changed their minds.

DOUBLE TROUBLE: BREASTS

Can't we have anything nice without people ruining it?

FACT 713 **The most popular bra size in the UK is 36D**, up from a 34B ten years ago. The increase is attributed to the popularity of cosmetic surgery, birth control pills and better nourishment. And wishful thinking.

FACT 714 An estimated 80 percent of women **wear the wrong bra size**, which can cause back pain, indigestion and headaches.

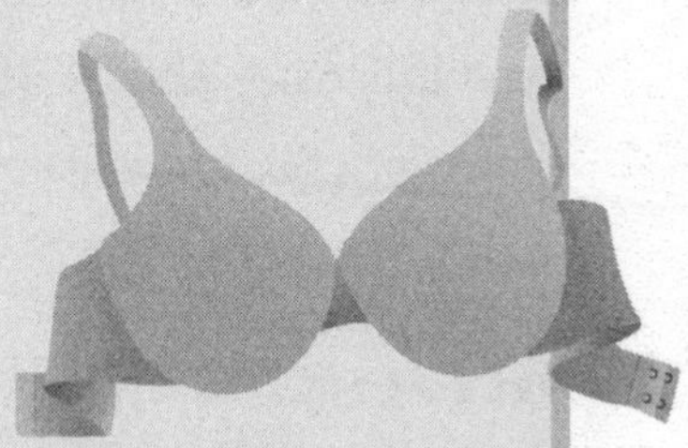

FACT 715 The mammary glands of a woman in her twenties are comprised of fat, milk glands and collagen. **As women age, the collagen and milk glands are replaced by more fat**, causing breasts to sag.

FACT 716 ☞ Men aren't the only ones with **hairy nipples**. Every woman has between two and fifteen dark, straight strands growing on her nipple at any time.

FACT 717 ☞ **Public toplessness is legal in five states**: Hawaii, Ohio, New York, Maine and—believe it or not—Texas, though women in the Lone Star State can still be arrested as a public nuisance.

FACT 718 ☞ Don't sunbathe topless on a beach in Dubai unless you want to **spend six months in jail**.

FACT 719 ☞ Exercising without a sports bra can give you **jogger's nipple**, which is soreness caused by the nipple rubbing against fabric. But you'll be the most popular sore-nippled jogger in your neighborhood.

FACT 720 ☞ **Sleeping facedown won't shrink your breasts**, but it can change their shape over time.

FACT 721 ☞ **One way to avoid saggy breasts is to stop smoking**. The chemicals in cigarettes break down the skin's elastin, which prevents sag.

FACT 722 ☞ A study presented in 2007 to the American Society of Plastic Surgeons found that **the main factors affecting breast sag** were age, smoking habits and the number of pregnancies a woman has had.

FACT 723 ☞ In 2010, a twenty-seven-year-old UK woman **almost suffocated her boyfriend with her size 40LL breasts**. He fell unconscious after being smothered by her boobs during sex.

FACT 724 ☞ Breasts are known to sag **during times of tension and stress**. "Is everything okay? Your breasts look a bit saggy today."

FACT 725 ☞ Though exercise can tone your pectoral muscles, **it does little to keep your breasts firm**, since they contain no muscle.

FACT 726 Breast enlargement is **the most popular cosmetic procedure in the world**, with more than three hundred thousand surgeries performed in the United States alone in 2008.

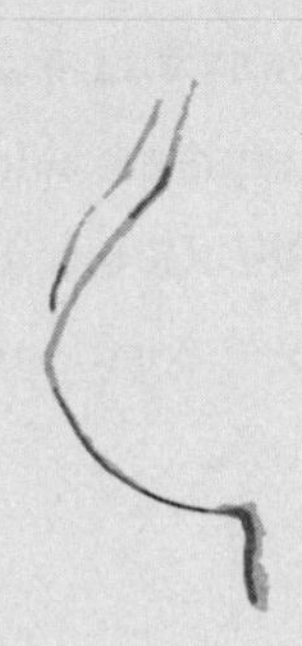

FACT 727 Almost twenty-one thousand women **had their implants removed** the same year.

FACT 728 Nearly **eighteen thousand breast reductions** were done on men in 2008.

FACT 729 Before breast implants were perfected, **surgeons tried numerous other materials**, including ivory, glass (?!), ground rubber, ox cartilage and silicone injections.

FACT 730 Popular in the 1960s, silicone injections to enlarge breasts **caused severe health problems for many women**, including tumors, granulomas and disfigurement.

FACT 731 ☞ **The world's largest breast implants** belong to Sheyla Hershey, who wears a size 38KKK bra. Each of Hershey's implants is 10,000cc, or 2.6 gallons, which is why boobs that big are called *jugs*.

FACT 732 ☞ Hershey has had **more than thirty cosmetic procedures** on her breasts.

FACT 733 ☞ **A man's nipple can lactate** if he's undergoing cancer treatments that require estrogen. Or if he watches *The Notebook*.

FACT 734 ☞ In the Catholic Church, **St. Agatha is the patron saint of breast cancer**, sterility, nurses and wet nurses, among others. Legend holds that she was tortured and had her breasts cut off before being martyred in AD 251.

FACT 735 ☞ Morgellons disease is a **mysterious skin disorder characterized by disfiguring sores and crawling or stinging sensations** on and under the skin.

FACT 736 ☞ **Morgellons sufferers** also report rashes, threads or black speck-like materials on or beneath the skin, fatigue, mental confusion, short-term memory loss, joint pain and changes in vision.

FACT 737 ☞ **Infants with the rare congenital condition known as macrodactyly have toes or fingers that are abnormally large due to the overgrowth of the underlying bone and soft tissue.**

FACT 738 ☞ Maple syrup urine disease (MSUD) is a genetic metabolism disorder in which the body cannot break down certain parts of proteins, **causing the victim's urine to smell like maple syrup**.

FACT 739 ☞ In times of physical stress such as infection or fever, MSUD **can damage the brain**.

FACT 740 ☞ **Jumping Frenchmen of Maine** is a rare disorder characterized by an unusually extreme startle reaction to a sudden or unexpected noise or sight. It was first identified in the late nineteenth century among an isolated population of French Canadian lumberjacks in Maine and Quebec.

FACT 741 ☞ **Dengue fever, a virus-based disease spread by mosquitoes, is a leading cause of illness and death in the tropics and subtropics.**

FACT 742 ☞ More than one-third of the world's population lives in areas of transmission of dengue fever. **As many as 100 million people are infected each year**.

FACT 743 ☞ **Symptoms of infection with the dengue virus** include sudden high fever, measles-like rashes, skin sensitivity, vomiting, headache, fatigue and low blood pressure.

FACT 744 ☞ Dengue fever is also characterized by acute joint and muscle aches, hence its nicknames, **"break-bone fever" and "bonecrusher disease."**

FACT 745 ☞ Alice in Wonderland syndrome **distorts visual perception** so that nearby objects appear disproportionately tiny, as though viewed through the wrong end of a telescope.

FACT 746 ☞ Alice in Wonderland syndrome is **frequently associated with migraines**.

FACT 747 ☞ Victims of pica exhibit a **compulsive appetite for non-edible items like clay, rocks, ash, paint, glue and even hair**. The condition is thought to be caused by deficient nutrition.

FACT 748 ☞ Pica is found among **the mentally ill, children and pregnant women living in poverty**.

FACT 749 ☞ Ondine's curse, or congenital central alveolar hypoventilation syndrome, is a **sleep disorder that leaves victims unable to breathe spontaneously**, so they must consciously will each breath or risk suffocating in their sleep.

FACT 750 ☞ Polydactylism is a congenital abnormality involving **being born with too many digits**, ranging from small nubs to fully formed fingers or toes.

FACT 751 ☞ While rare, polydactylism is prevalent **among communities known for intermarriage**, such as Philadelphia's Old Order Amish.

FACT 752 ☞ **People with Riley-Day syndrome** have an insensitivity to pain, which makes them highly accident-prone since they don't register warning signs of tissue damage like wounds or burns.

FACT 753 ☞ Half of Riley-Day patients

die from their injuries by age thirty.

FACT 754 ☞ Jerusalem syndrome is a **religious psychosis triggered by a visit to Jerusalem**, observed since medieval times. Its victims are overcome with religious fervor and believe they are prophets or messengers of God compelled to spread the gospel and encourage sinners to repent.

FACT 755 ☞ In 1969, an Australian man suffering from Jerusalem syndrome **set fire to the al-Aqsa Mosque on the Temple Mount**, convinced he was the "Lord's emissary." Widespread rioting ensued.

FACT 756 Hypertrichosis, **also called werewolf syndrome**, is a congenital condition that causes hair to grow all over the body like an animal's.

FACT 757 **One famous hypertrichosis victim was JoJo the Dog-Faced Boy**, aka Fedor Jeftichew, a Russian recruited by showman P. T. Barnum to tour with his circus in the nineteenth century.

FACT 758 Koro is one of a number of names for a hysterical condition known medically as **genital retraction syndrome**, whose victims become convinced that their genitals are disappearing into their bodies.

FACT 759 Koro syndrome can be contagious, **sparking off "penis panics"** like the one that overtook Singapore in 1967, when thousands of men became convinced that their penises were being stolen. *Penis Panic*—wasn't that a Village People album?

FACT 760 ☞ Named after the Greek god famous for changing his shape, Proteus syndrome is a rare progressive disorder that causes **disfiguring tumors and abnormal bone development**.

FACT 761 ☞ The most celebrated victim of Proteus syndrome was **Joseph Merrick, aka "The Elephant Man,"** a grotesquely deformed man in nineteenth-century London.

FACT 762 ☞ **Victims of Stendhal syndrome** experience dizziness, rapid heartbeat, confusion and hallucinations when exposed to large amounts of beautiful artwork.

FACT 763 ☞ Moebius syndrome is a rare neurological disorder that leaves victims expressionless because they **cannot move their faces**. They are often seen sleeping with their eyes open.

FACT 764 Progeria is a rare genetic condition that **causes young children to grow old prematurely**. The average child born with progeria will not live past age thirteen.

FACT 765 **Symptoms of progeria include a narrow, wrinkled face; hair loss; limited growth and short stature; oversized head; limited range of motion; heart disease; thinning bones; and arthritis.**

FACT 766 Victims of polyglandular Addison's disease **can literally die from fear or shock**, as their bodies are unable to produce adrenaline in response to alarm, but instead go into shock and begin to shut down vital organs.

FACT 767 ☞ Fournier's gangrene is a specific form of **necrotizing fasciitis (flesh-eating disease) that attacks the soft tissues of the genitalia**.

FACT 768 ☞ Though Fournier's **primarily affects men**, it can also develop in women due to episiotomies, abortions or hysterectomies.

FACT 769 ☞ Vulvar myiasis is a **parasitic infestation of a female's vulva caused by fly larvae**, which penetrate the skin and create painful open sores.

FACT 770 ☞ **Vulvar myiasis occurs predominantly in rural areas and is associated with poor hygiene.**

FACT 771 ☞ Paraneoplastic pemphigus (PNP) is a **rare autoimmune disease** that affects skin and/or mucous membranes with painful blisters.

FACT 772 ☞ Because PNP **typically occurs in association with cancer**, it is frequently fatal: 90 percent of victims will die due to sepsis, multi-organ failure or the cancer that caused the disease.

FACT 773 ☞ Fields's disease is a neuromuscular condition said to be **the rarest disease in the world**. Named after its only victims, Welsh twins Catherine and Kirstie Fields, the disease causes slow muscle deterioration and loss of movement.

FACT 774 ☞ Sufferers of the neurological disorder prosopagnosia, also called face blindness or facial agnosia, **cannot recognize familiar faces, including their own**. The illness can result from stroke, traumatic brain injury or other neurodegenerative diseases.

FACT 775 ☞ Some degree of prosopagnosia is often **present in children with autism and Asperger's syndrome** and may be the cause of their impaired social development.

FACT 776 ☞ Kleine-Levin syndrome is a rare disorder that primarily affects adolescent males and is characterized by **recurring periods of excessive sleep,** in some cases up to twenty hours per day.

FACT 777 ☞ Macular degeneration is one known cause of **Charles Bonnet syndrome**, which is characterized by hallucinations following a sudden change in vision.

FACT 778 ☞ The condition **may be aggravated by other circumstances** such as sensory deprivation (e.g., from living alone), diminished cognitive abilities, stroke, aging, depression or bereavement (e.g., "seeing" a deceased spouse).

FACT 779 ☞ A male baby **born with two penises** is said to have diphallia, or penile duplication. Victims of this very rare disorder will also typically have other congenital defects, and carry a higher risk for infection and death.

FACT 780 ☞ **The Fregoli delusion, or delusion of doubles**, is marked by the improbable belief that different people are in fact a single person who changes appearance or is in disguise.

FACT 781 ☞ A severe form of culture shock, **Paris syndrome is a condition exclusive to Japanese tourists** which causes them to need psychological treatment after visiting Paris and experiencing a city quite different from their expectations.

FACT 782 ☞ **Past victims of Paris syndrome** include two women who thought their hotel room was being bugged, a man who became convinced he was King Louis XIV, and a woman who believed she was being attacked with microwaves.

FACT 783 ☞ Reduplicative paramnesia is the **delusional belief that a location has been duplicated, existing in two or more places simultaneously**, or that it has moved to another site. For example, a person may believe that he is not in the hospital to which he was admitted, but an identical-looking hospital in a different part of the country, despite this being obviously false.

FACT 784 ☞ Though rare, **reduplicative paramnesia is most commonly associated with acquired brain injury**, particularly simultaneous damage to the right cerebral hemisphere and to both frontal lobes.

FACT 785 ☞ In a study of America's fifty largest metro areas, **Detroit was named our most stressful city**.

FACT 786 **Stress is strongly associated** with cardiac disease, hypertension, inflammatory diseases and compromised immune systems, and possibly to cancer.

FACT 787 Stress can break your heart, literally. Takotsubo cardiomyopathy, or "broken heart syndrome," occurs when the bottom of the heart balloons as **grief or other extreme stressors flood it with hormones**.

FACT 788 Stress **causes elevated levels of the hormone cortisol**, which gives us a short-term boost but also suppresses the immune system, elevates blood sugar and impedes bone formation.

FACT 789 Researchers at the University of California at San Francisco found an **association between high cortisol during late pregnancy in mothers and lower IQs in their children** at age seven.

FACT 790 Stress during pregnancy has also been **linked to children with autism**.

FACT 791 Researchers at Texas A&M found that **playing violent video games lowered, not increased, stress levels** in subjects.

FACT 792 Studies show that the hippocampus, an area of the brain central to learning and memory, is **damaged by long-term exposure to cortisol.**

FACT 793 **Chronic stress decreases your immune system's response to infection** and can affect your response to immunizations.

FACT 794 Studies show that HIV-infected men under high stress **are more likely to progress to AIDS** than those with lower levels of stress.

FACT 795 Chronic stress can **impair developmental growth** in children.

FACT 796 ☛ The stress of caring for a disabled spouse **increases your risk of stroke** significantly.

FACT 797 ☛ **Men are more likely than women** to develop stress-related disorders like hypertension, aggressive behavior, and abuse of alcohol and drugs.

FACT 798 ☛ The stress hormone cortisol not only causes abdominal fat to accumulate, but also enlarges individual fat cells, **leading to what researchers call "diseased" fat**.

FACT 799 ☛ A 2009 CNN poll revealed that the **top reason for stress in most countries is money**. Or, specfically, lack thereof.

FACT 800 ☛ Acoustic stress (caused by loud noises) can trigger an episode of **long QT syndrome (LQTS), a disorder of the heart's electrical system**. LQTS is estimated to cause as many as three thousand deaths in the United States every year.

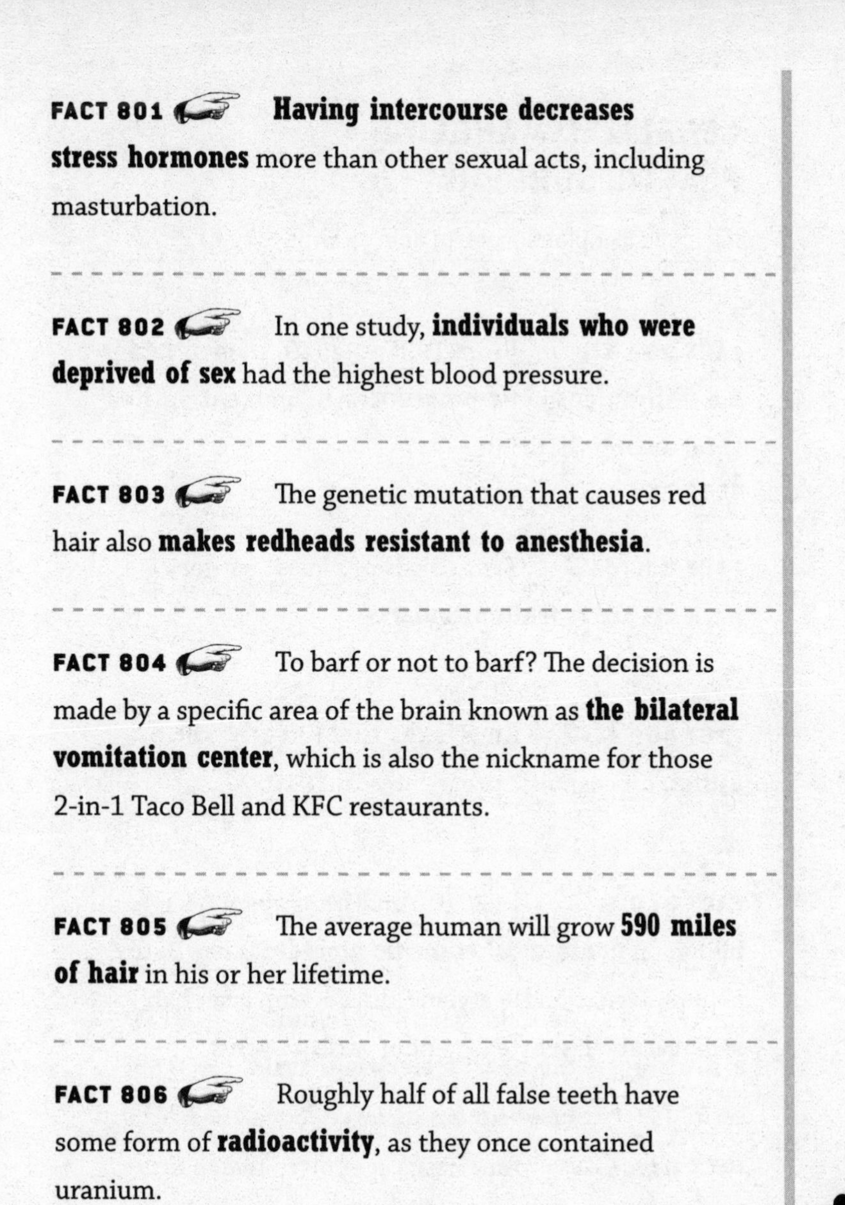

FACT 801 **Having intercourse decreases stress hormones** more than other sexual acts, including masturbation.

FACT 802 In one study, **individuals who were deprived of sex** had the highest blood pressure.

FACT 803 The genetic mutation that causes red hair also **makes redheads resistant to anesthesia**.

FACT 804 To barf or not to barf? The decision is made by a specific area of the brain known as **the bilateral vomitation center**, which is also the nickname for those 2-in-1 Taco Bell and KFC restaurants.

FACT 805 The average human will grow **590 miles of hair** in his or her lifetime.

FACT 806 Roughly half of all false teeth have some form of **radioactivity**, as they once contained uranium.

LEGALIZED MANGLING: PLASTIC SURGERY

If it looks like plastic, you're doing it wrong.

FACT 807 ☞ From 1997 to 2010, **women had 8.6 million cosmetic procedures**, 92 percent of the total during that time.

FACT 808 ☞ Two-thirds of plastic surgery patients are **repeat customers**.

FACT 809 ☞ **Americans spent nearly $10.7 billion** on cosmetic procedures in 2010.

FACT 810 ☞ In 2010, Americans spent $4.1 billion on **nonsurgical cosmetic procedures** like Botox, skin rejuvenation (Restylane, Juvéderm), laser hair removal and laser treatment of varicose veins.

FACT 811 ☞ Americans age thirty-five to fifty have **more cosmetic procedures than any other group**.

FACT 812 ☞ **Animals can get plastic surgery, too.** Procedures like eyelifts, facelifts, nose jobs and tummy tucks have been performed on canine breeds like pugs and bulldogs to alleviate skin folds and breathing problems. Or you can just spend about $50 and have them put down, which is a lot easier.

FACT 813 ☞ Other purely cosmetic procedures than can be done on pets are **testicular implants, cosmetic dentistry, Botox and breast reductions.**

FACT 814 ☞ The **top five countries with the most plastic surgeries** per capita are Switzerland, Cyprus, Spain, Lebanon and Greece. The United States ranks nineteenth.

FACT 815 ☞ A **voice lift** is a cosmetic procedure in which fat is implanted into the vocal cords to make the voice sound more youthful.

FACT 816 **Risks of voice lift surgery** include a voice that is permanently raspy, too soft or too high-pitched.

FACT 817 **Lesser-known cosmetic procedures** include knee lift, calf implants, ankle liposuction, toe tuck or shortening, pectoral (chest) implants, navel surgery, hymen reconstruction and dimple fabrication. If you have money to spend to shorten your toes, I hate you.

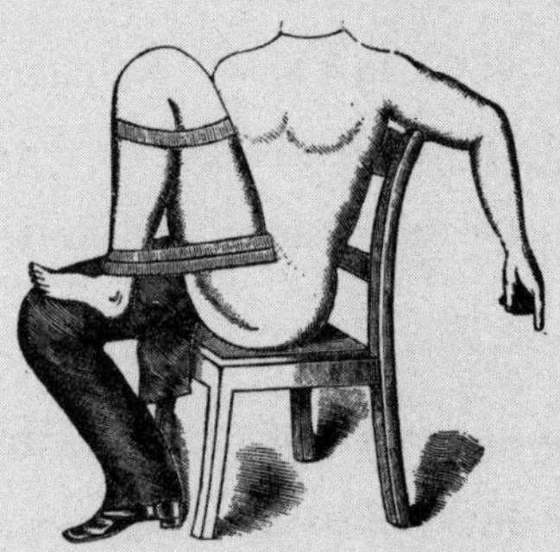

FACT 818 Potential complications of plastic surgery include **excessive bleeding, fluid buildup, infection and death**—the worst complication of all.

FACT 819 Loss of sensation in the nipple after a cosmetic breast procedure can occur **up to 70 percent of the time.**

FACT 820 Breast reduction surgery is **most likely to cause nipple numbness**, since the nipple and areola must be completely removed and reattached as skin grafts.

FACT 821 **Necrosis, or tissue death**, from surgical manipulation, is normal after most surgical procedures.

FACT 822 Your risk of necrosis **increases dramatically if you're a smoker**, as smoking affects blood supply to the tissues.

FACT 823 **Hematoma (excessive bleeding)** is another frequent post–cosmetic surgery occurrence.

FACT 824 Though rare, infection can result from plastic surgery and must be stopped immediately, as it **can spread very quickly**.

FACT 825 Liposuction can recontour body shape but **doesn't result in significant weight loss.**

FACT 826 Botox is a brand name of injectable botulinum toxin type A (BTX-A). In large amounts, **this toxin can cause botulism, a potentially fatal disease** associated with food poisoning.

FACT 827 Some patients who get repeated surgeries on the same feature suffer from a psychological condition called **body dysmorphic disorder** (BDD), or "imagined ugliness syndrome."

FACT 828 One in every fifty-one thousand plastic surgery patients who goes under the knife **will die.**

FACT 829 Combining surgeries **greatly increases the mortality risk**, as the patient must be under anesthesia for a longer period of time.

FACT 830 ☞ **Toxic levels of anesthetics** have been known to cause fatal respiratory failure in some patients. In rare instances, high levels of topical lidocaine have led to death.

FACT 831 ☞ In other cases, imitation botulinum toxin type A (BTX-A) or other imitation fillers not approved by the FDA have **left patients comatose or dead from surgery**.

FACT 832 The brain uses over **a quarter of the oxygen** inhaled by the human body.

FACT 833 Adult lungs have a surface area of almost seventy-seven yards—**more than three-quarters of a football field!**

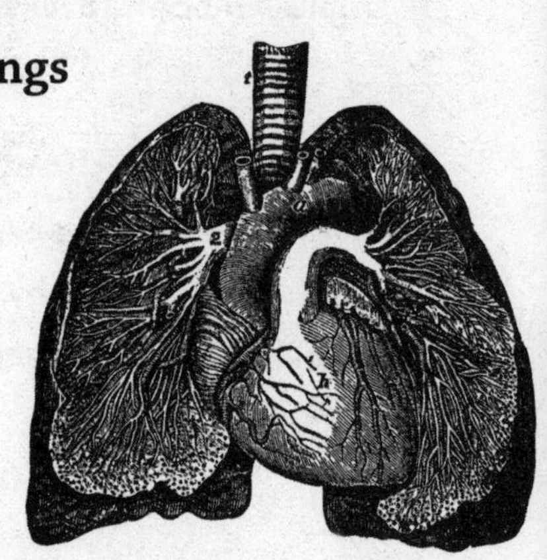

FACT 834 Your sense of smell is **ten thousand times more sensitive** than your sense of taste.

MAKING THE BEAST

The Naked Truth About Sex

THERE'S NOTHING SCARY ABOUT sex, nothing at all. That is, until you wake up naked and facedown in a stranger's bed, bald, blindfolded, gagged and handcuffed to the headboard, bleeding from at least two orifices and your newly pierced nipples, covered in leeches and dog bites and cigarette burns and missing part of an earlobe and a vital organ, with no recollection whatsoever of how you got there or why you smell like honey and have foreign currency and a note that reads, "Call me!" stuck to your ass.

FACT 835 During pregnancy, **uterine volume expands** from about one milliliter of fluid area to almost twenty liters.

FACT 836 **The womb shrinks back to half its pregnant weight within a week of giving birth. The ass takes a little longer.**

FACT 837 The word "taboo" is possibly derived from the Polynesian word for **menstruation**.

FACT 838 History's most prolific mother was an eighteenth-century Russian who gave birth to sixty-nine children in forty years: sixteen pairs of twins, seven sets of triplets and four sets of quadruplets. For forty years, **every time she stood up, a baby fell out of her vagina**.

FACT 839 ☞ **Vaginal prolapse occurs when the rectum, uterus, bladder or weakened pelvic muscles push on the vagina. In severe cases, the vagina can protrude from the body.**

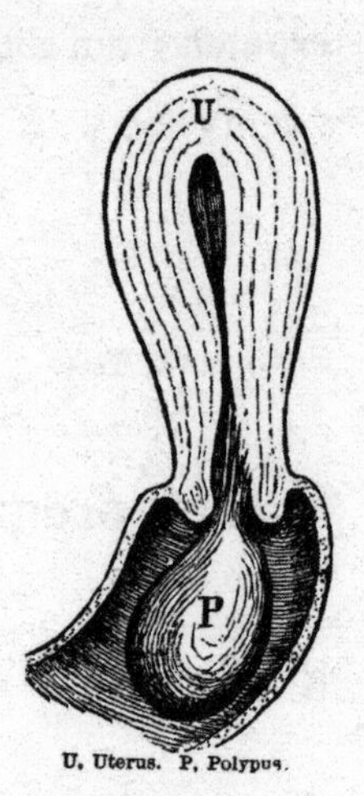

U. Uterus. P. Polypus.

FACT 840 ☞ **Vaginal dryness is a frequent symptom of menopause**, as is vaginal atrophy, when the organ becomes thinner and less elastic.

FACT 841 ☞ The clitoris contains **eight thousand nerve endings**, twice as many as the penis.

FACT 842 ☞ The average vagina is three to four inches long but can **expand by 200 percent** when sexually aroused. It's an optimistic organ.

FACT 843 ☞ Both vaginas and shark livers contain **squalene**, a natural lubricant.

FACT 844 ☞ Only a third of women **have orgasms from intercourse alone**.

FACT 845 ☞ Orgasms can lower a woman's risk of **heart disease, stroke, breast cancer and depression**.

FACT 846 ☞ **Orgasms can also strengthen a woman's immune system**, improve her sleep, regulate menstrual cycles, relieve menstrual cramps, lower stress levels and improve self esteem.

FACT 847 ☞ **Pregnancy can cause vaginal discharge, an increase in odors, itching and a swelling of the vulva that has been dubbed "cheeseburger crotch."**

FACT 848 Uterine rupture during pregnancy occurs most often at the site of a previous C-section incision; **it is a catastrophic complication** with a high incidence of fetal and maternal morbidity.

FACT 849 **The only mammals that experience menopause are elephants, humpback whales and human females.**

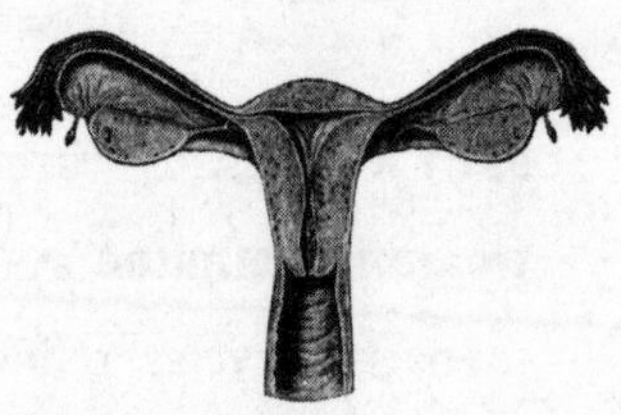

FACT 850 **Fourth-degree vaginal tears** during pregnancy can damage the perineal muscles, the sphincter and the lining of the rectum.

FACT 851 Toxic shock syndrome (TSS) is a rare, bacteria-caused illness occurring mostly in **menstruating women who use high-absorbency tampons**. Symptoms include vomiting, diarrhea, high fever and drops in blood pressure.

FACT 852 In severe cases, TSS infection can cause **kidney and liver failure**.

FACT 853 **You can get pregnant during menstruation**: sperm can live inside a woman for three to five days after intercourse, and ovulation can occur during or shortly after the bleeding phase.

FACT 854 Over her lifetime, **the average woman will spend roughly 3,500 days menstruating** and use about 11,400 tampons.

FACT 855 Women's periods tend to be heavier, more painful and longer **during cold months**.

FACT 856 In some parts of India, a woman indicates that she is menstruating by **wearing a handkerchief around her neck stained with her menstrual blood.**

FACT 857 ☞ Some scientists have suggested **harvesting stem cells** from menstrual blood.

FACT 858 ☞ The clitoris is the only organ in the human body that **exists solely for sexual pleasure**.

FACT 859 ☞ **Clitoraid is a nonprofit organization whose goal is to repair the clitorises of victims of female circumcision, not a sports drink for women.**

FACT 860 ☞ The average clitoris is around four inches in length, but, like a penis, the **majority of the organ** is hidden from view within the body.

FETISHES: SEX OUTSIDE THE BOX. *WAY* OUTSIDE.

We don't mean to judge, but some of you need professional help.

FACT 861 ☞ The World Health Organization and psychiatry's DSM-IV manual classify "fetishism" as a **mental illness.**

FACT 862 ☞ The World Health Organization describes a fetish as a "**reliance on some non-living object as a stimulus for sexual arousal and sexual gratification**." The fetish object might "simply serve to enhance sexual excitement achieved in ordinary ways (e.g. having a partner wear a particular garment)."

FACT 863 ☞ Individuals with **agalmatophilia** are sexually attracted to mannequins, dolls or statues.

FACT 864 ☞ People who like to dress up in animal costumes for sexual gratification have ursusagalmatophilia, which nobody can pronounce, so they **call themselves "furries" or "plushies"** instead. Or "single."

FACT 865 ☞ Individuals with osmolagnia are **sexually aroused by bodily odors** like sweat and flatulence.

FACT 866 ☞ Paraphilic infantilism is sex play while **wearing diapers and pretending to be a baby**.

FACT 867 ☞ Women with hybristophilia are **attracted to "bad boy" types**, which can include prisoners serving time for horrific crimes.

FACT 868 ☞ A fetish known as **hematolagnia involves using or even drinking blood in a sexual way**. "With the popularity of *Twilight*, I've seen an increase in vampire fantasies and biting," says one sex therapist. Everyone else has seen an increase in suck.

FACT 869 Sexual arousal from biting is called **odaxelagnia**.

FACT 870 For people with mechanophilia, **motorized vehicles and gadgets** are central to the sexual experience.

FACT 871 **If tears turn you on**, you might be a dacryphiliac.

FACT 872 Teratophilia is **sexual attraction to deformed or monstrous people**; one form of this is acrotomophilia, sexual attraction to amputees.

FACT 873 Emetophilia involves vomiting or watching others vomit **for sexual pleasure**.

FACT 874 ☞ Crush fetishists enjoy watching small insects or animals **being crushed to death**. Inanimate objects like cigarettes, fruit or toy cars can also be crush fetish stimuli.

FACT 875 ☞ Klismaphilia is the fetish in which **pleasure is derived from enemas**. It is sometimes regarded as a form of anal masturbation.

FACT 876 ☞ Necrophilia is a sexual attraction to human corpses. This fetish **can lead to grave-robbing** and sexual activity with the dead body.

FACT 877 ☞ **Squashing** is a fetish that involves one person being squashed by a much heavier partner.

FACT 878 ☞ Balloon fetishists, or "looners," are people who are sexually stimulated by and/or have sex with balloons, **similar to those with latex fetish**.

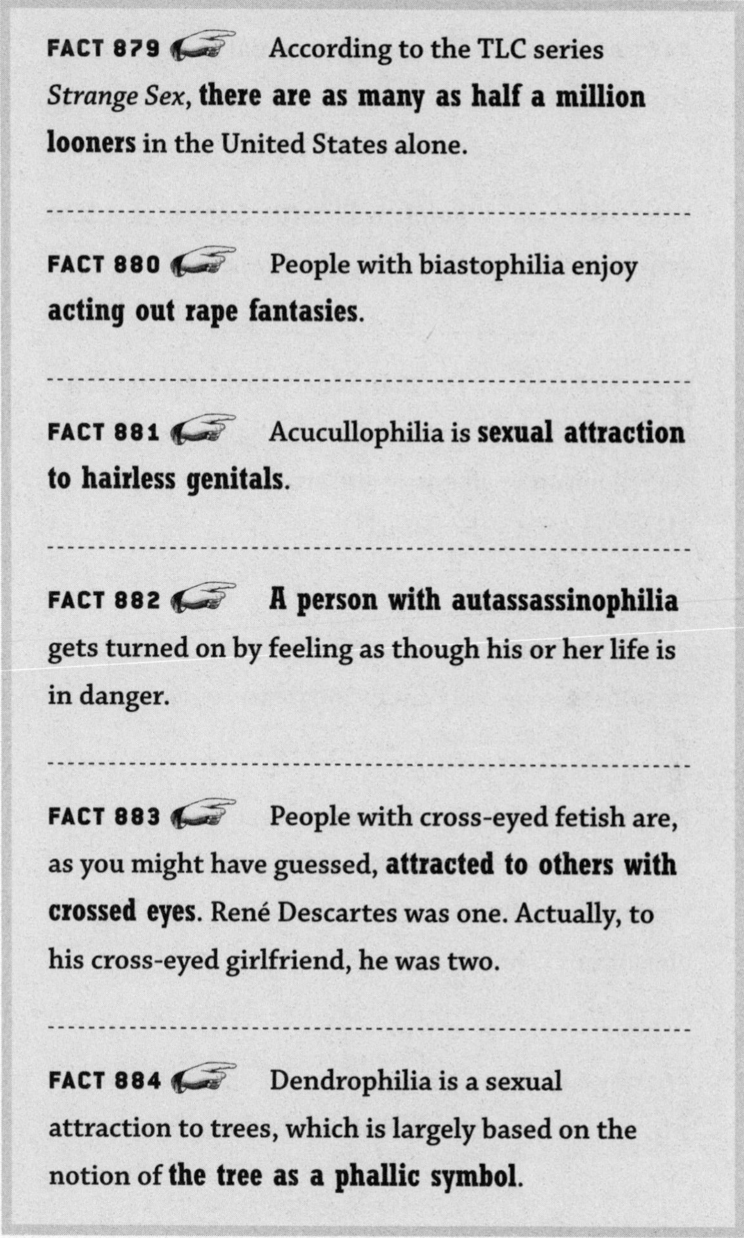

FACT 879 ☛ According to the TLC series *Strange Sex*, **there are as many as half a million looners** in the United States alone.

FACT 880 ☛ People with biastophilia enjoy **acting out rape fantasies**.

FACT 881 ☛ Acucullophilia is **sexual attraction to hairless genitals**.

FACT 882 ☛ **A person with autassassinophilia** gets turned on by feeling as though his or her life is in danger.

FACT 883 ☛ People with cross-eyed fetish are, as you might have guessed, **attracted to others with crossed eyes**. René Descartes was one. Actually, to his cross-eyed girlfriend, he was two.

FACT 884 ☛ Dendrophilia is a sexual attraction to trees, which is largely based on the notion of **the tree as a phallic symbol**.

FACT 885 ☞ Pyrophilia is sexual gratification from **fire or arson**.

FACT 886 ☞ **Retifism is better known as a foot fetish**, or sexual attraction to feet or shoes.

FACT 887 ☞ **For individuals with taphephilia**, sexual pleasure is derived from being buried alive. But then you're all horny and stuck in the ground alone, so what's the point?

FACT 888 ☞ Archnephilia is **sexual attraction to spiders**, especially daddy longlegs.

FACT 889 ☞ Frotteurism is sexual arousal by **nonconcensual groping or rubbing one's genitals against a victim**, typically in crowded places like an elevator or concert.

FACT 890 ☞ Tamakeri is a Japanese fetish where men love to be **kicked in the balls by women**.

FACT 891 Oculolinctus involves **licking someone's eyeball** for sexual pleasure.

FACT 892 **Chrematistophiliacs** get off on having to pay for sex or being robbed by a sexual partner.

FACT 893 Mysophilia is sexual gratification from **smelling or tasting things like sweaty underwear, soiled clothing or used tampons.**

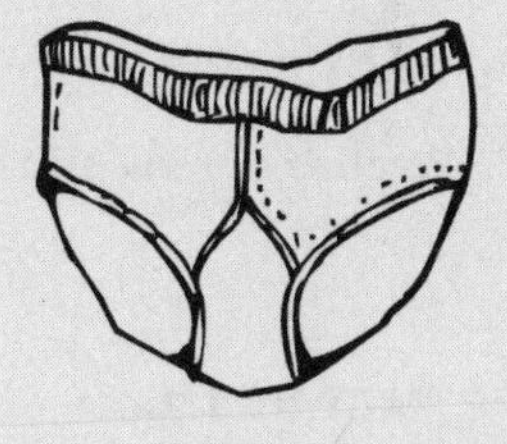

FACT 894 Symphorophilia is **sexual arousal from car accidents** or other disasters, as portrayed in the 1996 film *Crash* (not the 2005 Oscar winner for Best Picture).

FACT 895 Smoking can **shrink a man's penis** by up to a centimeter.

FACT 896 The longest recorded distance of a male ejaculation was **more than eighteen feet**. Who measures these things?

FACT 897 Sperm travel as fast as two hundred inches per second, **or about 11 mph**.

FACT 898 **Congenital hypoplasia** occurs when the penile glands are stuck directly to the pubic bone, resulting in a very small penis.

FACT 899 **Despite what men might claim, only 15 percent have a penis longer than seven inches. Only 3 percent have a penis more than eight inches long.**

FACT 900 **The world's largest recorded penis** belongs to forty-one-year-old New Yorker Jonah Falcon, whose appendage measures 9.5 inches flaccid and 13.5 inches erect.

FACT 901 Handyman Charles Lennon has "suffered" through a **perpetual erection since being fitted with a steel penis implant** in 1996. The good news is, he doesn't have to carry a hammer anymore.

FACT 902 Lennon sued the implant's maker, Dacomed Corp., over his decade-long boner and **was awarded $400,000 for pain and suffering**. He used some of the money to buy bigger pants.

FACT 903 In a Spanish study on male attractiveness, women consistently disregarded looks and chose men **who turned out to have the healthiest sperm**.

FACT 904 **One in every 400 men** is flexible enough to give himself a blow job. The other 399 have tried.

FACT 905 **The average male orgasm lasts six seconds; the average female orgasm lasts twenty seconds.**

FACT 906 The impulse to ejaculate **comes from the spinal cord**; no brain is needed.

FACT 907 The most common cause of penile rupture is **vigorous masturbation**. If that happens, you're doing it wrong.

FACT 908 Size might matter after all: a study done by the State University of New York found that **the longer your penis, the better your semen's chances of fertilizing a female ovum**, particularly in the presence of competing sperm.

FACT 909 History's busiest penis surely belonged to King Fatefehi of Tonga, who reportedly **deflowered 37,800 women between 1770 and 1784**, or about seven virgins a day. It's good to be king.

FACT 910 **There is no relation** between penis size and race.

FACT 911 Circumcised foreskin can be reconstructed, but the process requires the use of **plastic rings, caps and weights**.

FACT 912 If a man claims he lost control of his penis, he might be telling the truth. **The penis is controlled by his autonomic nervous system**, which also regulates heart rate and blood pressure, so sexual arousal is mostly involuntary.

FACT 913 **Heavy lifting or straining to have a bowel movement can also produce an erection.**

FACT 914 A man's penis is **shaped like a boomerang**: the root is tucked inside his pelvis and attached to his pubic bone.

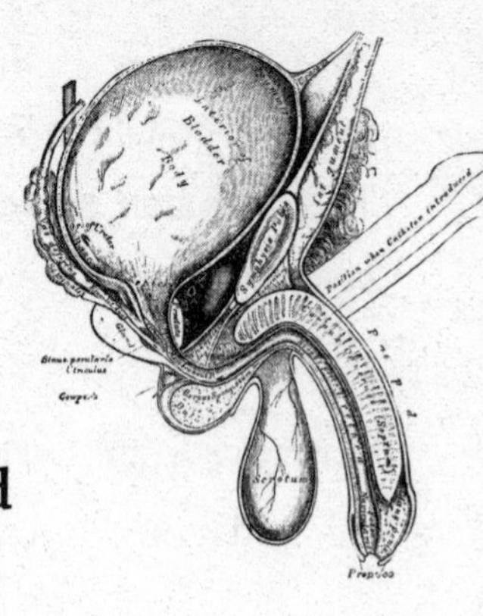

FACT 915 **One method of surgical penis enlargement** is to cut the suspensory ligament that holds the root of the penis up inside the pelvis.

FACT 916 Surgical penis enlargement **can only add an inch or so** to the length or your penis. But for a guy with a one-centimeter schlong, an extra inch is a big deal.

FACT 917 One team of researchers reports that while 85 percent of women are satisfied with their partner's penis size, **only 55 percent of men were happy with their own size**.

FACT 918 ☞ The same research team found that **0.2 percent of men** surveyed would prefer a smaller penis.

FACT 919 ☞ A 2002 study published in Europe found that **only 20 percent of women surveyed considered penile length "important."** Only 1 percent considered it "very important."

FACT 920 ☞ Ninety percent of female respondents in a 2001 poll said that the **thickness of a penis contributes more to their sexual pleasure than the length**.

FACT 921 ☞ The largest penis in the animal kingdom in relation to size belongs to the barnacle, whose **40:1 penis ratio** would give him a 240-foot-long organ if he were human. And he would still say it was 280.

FACT 922 ☞ In relative terms, the **tuberous bushcricket has the world's biggest testicles**—a full 14 percent of its body weight. The same ratio on a two-hundred-pound man would give him a twenty-eight-pound set of nuts. And a backache.

FACT 923 ☞ During fetal development, the gonads for males and females are the same until about six weeks of age. The sex organs **will become either testicles or ovaries** during the first trimester of pregnancy.

FACT 924 ☞ Testicles sweat for a reason: the **perspiration keeps the balls cool** to ensure better sperm production.

FACT 925 ☞ **The left testicle** usually hangs lower than the right for right-handed men. The opposite is true for lefties.

FACT 926 ☞ Undescended testes are **common in male babies**, affecting a third of premature infants and 4 percent of on-time newborns.

FACT 927 ☞ **Undescended testicles can cause infertility**, sexual dysfunction and a higher risk of testicular cancer.

FACT 928 Testicles are sensitive because **the body wants you to protect this vital reproductive area**.

FACT 929 The nerves in testicles are **connected to the abdomen,** another reason why force to the area causes so much pain, and why that pain is felt in the stomach.

FACT 930 **"Blue balls" is a real condition**. If a male is aroused for a long period of time and does not orgasm, fluid builds up and cramping occurs.

FACT 931 Contact sports, like football, soccer and martial arts, can result in **traumatic blows to the groin**, affecting a male's sperm production and fertility—and his desire to continue playing contact sports like football, soccer and martial arts.

FACT 932 Some blows to the groin can rupture the testicles, causing **bleeding in the scrotum**.

FACT 933 Orchitis is inflammation of one or both testicles and may be **caused by bacteria or a virus like mumps**.

FACT 934 Testicular cancer is most common in men **between the ages of twenty and thirty-nine**.

FACT 935 Factors that increase a man's **risk of testicular cancer** include age, family history of testicular cancer, delayed descent of testicles as an infant, and abnormal testes development.

FACT 936 A man's lifetime chance of having testicular cancer is about **1 in 270**.

FACT 937 As testicular cancer is **one of the most curable forms of cancer**, the risk of death is low, at about 1 in 5,000.

FACT 938 The American Cancer Society estimates that in 2012 there will be about **8,590 new cases of testicular cancer** in the United States.

FACT 939 Three hundred and sixty of those cases will be **fatal**.

FACT 940 Hitler and Napoleon each had **only one testicle**. Together they were two nuts.

FACT 941 At five hundred kilograms (eleven hundred pounds) each, **the testes of the right whale are the largest in the animal kingdom**.

FACT 942 An octopus's testicles are located **in its head**.

FACT 943 During mating with the queen bee, the **male drone's genitals explode** and snap off inside the queen.

FACT 944 **The word "avocado"** comes from the Aztec word for testicles.

FACT 945 **The orchid's name is based on the Greek word *órkhis*, literally meaning "testicle," because of the shape of its root.**

FACT 946 There are more than fifteen thousand videos **listed under "hit in the balls"** on YouTube.

FACT 947 Almost **three-quarters of men age seventy** are still potent.

FACT 948 ☞ In a **2007 MSNBC poll**, 28 percent of married men admitted to having extramarital sex, compared to 18 percent of married women.

FACT 949 ☞ The same poll found that 15 percent of men and 7 percent of women have **engaged in online sex or sexual webcamming**.

FACT 950 ☞ Around 90 percent of the total male population say they **masturbate**, but only 60 to 65 percent of the total female population admit it.

FACT 951 ☞ This difference might be explained by the fact that **many women don't begin masturbating until their twenties or thirties**, while men usually begin in their early teens.

FACT 952 ☞ **Men lie more often than women about the number of sex partners they have had.**

FACT 953 In a *Cosmopolitan* poll, only 25 percent of women and 12 percent of men **reported being completely satisfied with their sex lives**.

FACT 954 Fewer than 10 percent of women say they've participated in a threesome; **14 percent of men** say they have.

FACT 955 A third of men polled say that **having a threesome with their partner** is their top sexual fantasy, but a quarter of men have no interest in a threesome.

FACT 956 Almost 10 percent of women polled say they have **masturbated at work**. Almost 10 percent of men say they have masturbated *on* work.

FACT 957 More men than women admit **accessing porn at work**, but not by much: 20 percent of men vs. 13 percent of women.

STDs: NATURE'S CHASTITY BELT

Because every good thing has a catch.

FACT 958 ☞ The World Health Organization (WHO) estimates that **340 million cases of curable STDs** occur every year.

FACT 959 ☞ Many sexually transmitted diseases **can remain asymptomatic** for years.

FACT 960 ☞ Chlamydia is the **most frequently reported bacterial sexually transmitted disease** in the United States, with an estimated 2.8 million infections every year.

FACT 961 ☞ Because symptoms of chlamydia in women are usually mild or absent, **serious complications that cause irreversible damage can occur** before a woman ever recognizes a problem.

FACT 962 ☞ The CDC estimates that more than seven hundred thousand people in the United States **contract gonorrhea** every year.

FACT 963 ☞ Gonorrhea can affect the **anus, eyes, mouth, genitals or throat**.

FACT 964 ☞ **Pelvic inflammatory disease**, a serious complication of STDs like chlamydia and gonorrhea, can lead to internal abscesses—pus-filled pockets that are hard to cure—and chronic pelvic pain.

FACT 965 ☞ Gonorrhea can be **life-threatening** if it spreads to the blood or joints.

FACT 966 ☞ Drug-resistant strains of gonorrhea are **increasing in many areas of the world**, including the United States.

FACT 967 ☞ **The merkin, or pubic wig**, became popular in the 1400s among prostitutes who shaved off their pubic hair due to lice infection or who needed to cover up sores and other evidence of STDs.

FACT 968 ☞ Hepatitis A is an **acute liver disease** acquired from ingestion of even microscopic amounts of fecal matter.

FACT 969 ☞ **Hepatitis B, which can lead to liver disease or liver cancer**, is acquired by contact with infectious blood, semen or other body fluids of an infected person.

FACT 970 ☞ You can get Hepatitis C from contact with the blood of an infected person, primarily through **sharing contaminated needles to inject drugs**. The virus can lead to cirrhosis of the liver and liver cancer.

FACT 971 ☞ **One out of every six Americans** aged fourteen to forty-nine has a genital herpes HSV-2 infection.

FACT 972 ☞ Visible open sores are not required to contract herpes (HSV-1 and HSV-2); **transmission can occur from a partner who has no evidence of infection** and may not know that he or she has the disease.

FACT 973 ☞ **Herpes may play a role in the spread of HIV**: herpes can make people more susceptible to HIV infection and can make HIV-positive individuals more infectious.

FACT 974 ☞ At least **half of sexually active men and women** will get HPV (human papillomavirus) at some point in their lives.

FACT 975 ☞ **Untreated HPV infections** can cause cancer of the cervix, vulva, vagina, penis, anus or oropharynx (back of the throat, including the base of the tongue and the tonsils).

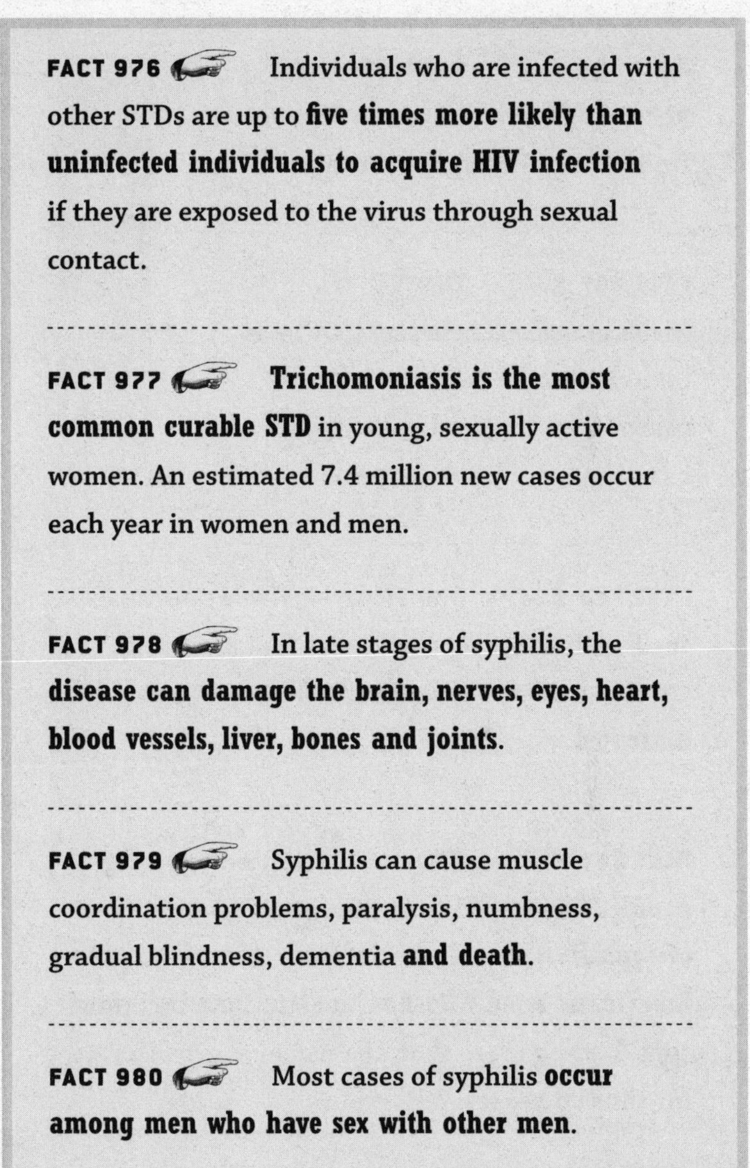

FACT 976 Individuals who are infected with other STDs are up to **five times more likely than uninfected individuals to acquire HIV infection** if they are exposed to the virus through sexual contact.

FACT 977 **Trichomoniasis is the most common curable STD** in young, sexually active women. An estimated 7.4 million new cases occur each year in women and men.

FACT 978 In late stages of syphilis, the **disease can damage the brain, nerves, eyes, heart, blood vessels, liver, bones and joints**.

FACT 979 Syphilis can cause muscle coordination problems, paralysis, numbness, gradual blindness, dementia **and death**.

FACT 980 Most cases of syphilis **occur among men who have sex with other men**.

FACT 981 ☞ **Condoms do not prevent STDs**: you can catch diseases from skin-on-skin genital contact.

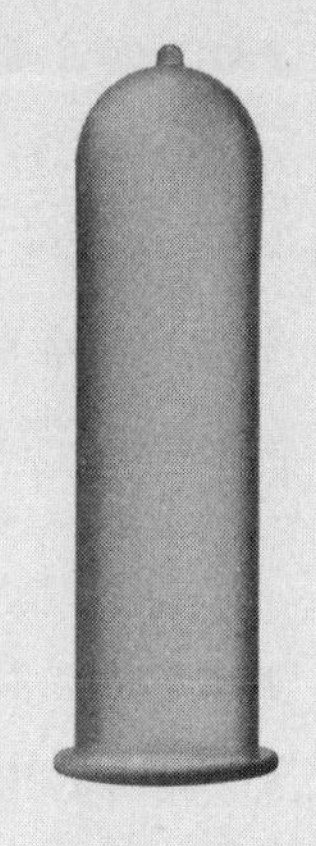

FACT 982 ☞ The STD *lymphogranuloma venereum* (LGV) was once rare but is now **becoming more common in developed nations**, and can be fatal.

FACT 983 ☞ *Granuloma inguinale* causes small, painless ulcers and warts that burst and spread, making it **a dangerous STD infection if left untreated**.

FACT 984 ☞ STDs aren't limited to young people. From 2005 to 2009, the **number of syphilis and chlamydia cases among Americans aged fifty-five to sixty-four increased significantly** more than the national growth rate for those diseases.

FACT 985 ☞ In the Middle Ages, **syphilis was treated with mercury**, a highly toxic heavy metal that occasionally killed the patient. That's one way to cure the disease.

FACT 986 ☞ Herpes has been around since ancient times. Roman emperor **Tiberius** banned kissing at public events to curb the spread of the disease.

FACT 987 ☞ Herpes was once thought to be **caused by insect bites**.

FACT 988 ☞ Syphilis infection of an unborn fetus by its mother can result in **spontaneous abortion, stillbirth, or death in the mother's womb**. Birth defects are common for those babies who survive delivery.

FACT 989 ☞ In July 1981, the *New York Times* reported an outbreak of a rare form of cancer among gay men in New York and California after emergency rooms began to see a rash of seemingly healthy young men with fevers, flu-like symptoms and pneumonia. This was the beginning of what has become **the biggest health care concern in modern history**, the HIV/AIDS epidemic.

FACT 990 ☞ In 1984 there were eight thousand confirmed cases of HIV in the United States. By 1990 that number **had grown to 1 million.**

FACT 991 ☞ By the end of 2003, AIDS had **orphaned 12 million children in sub-Saharan Africa.**

FACT 992 ☞ An estimated **1.3 million Africans died from AIDS** in 2009.

FACT 993 The first name suggested for AIDS, **Gay-Related Immune Deficiency (GRID)**, was rejected because it did not account for at-risk individuals such as intravenous drug users, recipients of blood transfusions, and heterosexual partners of AIDS patients.

FACT 994 The 2009 rate of chlamydia infection for women was **three times that of men**.

FACT 995 The **2009 rate of syphilis for men** was more than five times that of women.

FACT 996 Alabama, Louisiana and Mississippi all placed among the top five states for syphilis, gonorrhea and chlamydia cases in 2009. Mississippi ranked number one in two of the three categories. Alaska was **the only non-Southern state to appear in any of the lists**.

FACT 997 ☞ Women who read romance novels have sex **twice as often** as those who don't.

FACT 998 ☞ After fingers and vibrators, **candles** are the phallic objects used most often by female masturbators. Unlit ones, hopefully.

FACT 999 ☞ When two people kiss, **they exchange between 10 million and 1 billion bacteria**.

FACT 1,000 ☞ **Kissing is good for teeth: the anticipation of a kiss increases the flow of saliva in the mouth, giving the teeth a plaque-dispersing bath.**

FACT 1,001 ☞ At **the 2009 Masturbate-A-Thon in Denmark,** a woman set the distance record by squirting her ejaculate an impressive 10.3 feet.

FACT 1,002 The **world record for ongoing masturbation** belongs to a man who pleasured himself for nearly ten straight hours at the San Francisco Masturbate-A-Thon in 2009.

FACT 1,003 For pain relief, **sex is ten times more effective than Valium.**

ACKNOWLEDGMENTS

I AM EXCEEDINGLY grateful to the people who helped bring this book to life: Holly Schmidt and Allan Penn at Hollan Publishing; John Duff and my brilliant, patient editor Meg Leder at Perigee Books; Cyndi Culpepper, Andrea Nay, Paul Markowski, Connie Biltz and Rachael Pavlik, whose wit and friendship inspire me; and my wife, Paige, and daughter, Keaton, whose love and support sustain me.

SOURCES

CELEBRITY FREAK CLUB

1 www.trutv.com/conspiracy/phenomena/urine/gallery .html?curPhoto=3

2 Ibid.

3 www.hollywoodauditions.com/Biographies/courtney_cox.htm

4 http://news.bbc.co.uk/2/hi/entertainment/6278145.stm

5 www.snopes.com/movies/actors/chaplin2.asp

6 http://omgfacts.com

7 www.guardian.co.uk/world/2010/apr/18/george-washington -library-new-york

8 www.cracked.com/article_16472_6-absurd-phobias-and-people -who-actually-have-them.html

9 www.ebizarre.com/Category/People/19/#190

10 *Entertainment Weekly*, September 13, 1999, www.ew.com/ew/ article/0,84652,00.html

11 http://listverse.com/2010/11/03/10-well-known-people-and -their-phobias/

12 www.tvguide.com/celebrities/patrick-dempsey/bio/169151

13 www.hollywood.com/news/Stewart_Stopped_Dating_Hopkins_ Because_of_Hannibal/3600693

14 http://ebizarre.com/Category/People/3/#28

15 http://abcnews.go.com/Entertainment/WolfFiles/ story?id=116591&page=1

16 www.uh.edu/engines/epi1642.htm

17 www.playboy.com/articles/lil-wayne-dirty-dozen/index .html?page=2

18 *Believe You Can: The Power of a Positive Attitude* by John Mason (Revell, 2010)

19 www.fbi.gov/about-us/history/famous-cases/al-capone

20 http://listverse.com/2010/11/03/10-well-known-people-and -their-phobias/

21 www.people.com/people/charlize_theron

22 www.grammy.com

23 *Last Train to Memphis: The Rise of Elvis Presley* by Peter Guralnick (Little, Brown, 1994)

24 Ibid.
25 Ibid.
26 Ibid.
27 *Rock 'n' Roll* by Dave Rogers (Routledge & Kegan Paul, 1982)
28 *The FBI Files on Elvis Presley* by Thomas Fensch (New Century Books, 2001)
29 *Last Train to Memphis: The Rise of Elvis Presley* by Peter Guralnick (Little, Brown, 1994)
30 *Elvis Day by Day: The Definitive Record of His Life and Music* by Peter Guralnick and Ernst Jorgensen (Ballantine, 1999)
31 *Elvis Presley: The Ed Sullivan Shows* by Greil Marcus, DVD booklet (Image Entertainment, 2006)
32 *The Columbia History of American Television* by Gary R. Edgerton (Columbia University Press, 2007)
33 *Elvis Presley: The Ed Sullivan Shows* by Greil Marcus, DVD booklet (Image Entertainment, 2006)
34 *The Elvis Encyclopedia* by Adam Victor (Overlook Duckworth, 2008)
35 *Elvis Day by Day: The Definitive Record of His Life and Music* by Peter Guralnick and Ernst Jorgensen (Ballantine, 1999)
36 *Last Train to Memphis: The Rise of Elvis Presley* by Peter Guralnick (Little, Brown, 1994)
37 "A Whole Lotta Elvis Is Goin' to the Small Screen" by Curt Fields, *Washington Post*, August 3, 2007
38 "Girls! Girls! Girls!" by Ruthe Stein, *San Francisco Chronicle*, August 3, 1997
39 *Elvis: The Biography* by Jerry Hopkins (Plexus, 2007)
40 *Last Train to Memphis: The Rise of Elvis Presley* by Peter Guralnick (Little, Brown, 1994)
41 www.nme.com/festivals/photos/75-things-you-might-not-know-about-elvis-presley/162384/77/1#77
42 *Elvis the #1 Hits: The Secret History of the Classics* by Patrick Humphries (Andrews McMeel, 2003)
43 "Doctor Feelgood" by Alan Higginbotham, *Observer*, August 11, 2002
44 *Elvis: The Biography* by Jerry Hopkins (Plexus, 2007)
45 *Down at the End of Lonely Street: The Life and Death of Elvis Presley* by Peter Harry Brown and Pat H. Broeske (Signet, 1997)
46 Ibid.
47 www.thedailybeast.com/newsweek/blogs/the-gaggle/2009/12/21/the-day-nixon-met-the-king.html
48 www.belfasttelegraph.co.uk/entertainment/music/news/elvis-30-weird-and-wonderful-facts-13467073.html#ixzz1TvqczRMd
49 *Careless Love: The Unmaking of Elvis Presley* by Peter Guralnick (Back Bay Books, 1999)
50 http://ebizarre.com/Category/Entertainment/16/ #157
51 *Oprah: A Biography* by Kitty Kelley (Crown Books, 2010)

52 www.thesun.co.uk/sol/homepage/features/2799410/75-facts-about-Elvis-Presley-for-his-75th.html
53 Ibid.
54 www.cracked.com/article_16472_6-absurd-phobias-and-people-who-actually-have-them.html
55 www.timesonline.co.uk/tol/life_and_style/men/article1820060.ece
56 www.digitalhistory.uh.edu/database/article_display.cfm?HHID=450
57 www.nytimes.com/1996/01/07/nyregion/richard-versalle-63-met-tenor-dies-after-fall-in-a-performance.html
58 "Rapper Ate Flesh for 'Gangsta Image,'" *Age Australia*, April 14, 2003
59 www.thefrisky.com/post/246-8-ladies-who-were-rejected-by-playboy/
60 www.mentalfloss.com/blogs/archives/19106
61 Ibid.
62 Ibid.
63 Ibid.
64 www.neatorama.com/2007/03/26/10-strange-facts-about-einstein/
65 Ibid.
66 www.thedailybeast.com/articles/2010/08/03/what-the-great-ate-famous-people-and-their-infamous-appetites.html
67 www.foxnews.com/slideshow/entertainment/2011/07/12/celebrity-secrets/
68 Ibid.
69 www.foxnews.com/slideshow/entertainment/2011/03/03/infamous-celebrity-meltdowns
70 http://94.236.123.156:8082/biographies/charlize-theron.html
71 www.cosmopolitan.com/celebrity/exclusive/leighton-meester-cover-interview
72 www.thedailybeast.com/newsweek/2007/10/01/the-darkest-secret.html
73 www.thesun.co.uk/sol/homepage/news/90628/Spideys-dads-bank-raid.html
74 ww.people.com/people/article/0,,20009251,00.html
75 www.bet.com/news/music/2012/04/11/nicki-minaj-irked-by-lady-gaga-comparison.html
76 www.people.com/people/archive/article/0,,20149383,00.html
77 www.mentalfloss.com/blogs/archives/19106
78 *Truth and Rumors: The Reality Behind TV's Most Famous Myths* by Bill Brioux (Greenwood Publishing, 2007)
79 Ibid.
80 www.imdb.com/name/nm0001820/bio
81 *Truth and Rumors: The Reality Behind TV's Most Famous Myths* by Bill Brioux (Greenwood Publishing, 2007)
82 http://AbeVigoda.com/Facebook
83 http://mashable.com/2009/10/13/zach-braff-video/
84 http://cm1.theinsider.com/news/38908_Celebrity_Obituary_Screw_

Ups/index.html
85 http://brainz.org/murdered-media-25-premature-celebrity-obituaries/
86 Ibid.
87 Ibid.
88 Ibid.
89 Ibid.
90 Ibid.
91 Ibid.
92 www.aoltv.com/2006/06/03/jaleel-white-is-not-dead/
93 www.staugustine.com/stories/062204/com_2404632.shtml
94 http://articles.chicagotribune.com/2011-12-30/news/ct-talk-twitter-not-dead-yet-1231-20111224_1_internet-rumors-twitter-web-censorship
95 Ibid.
96 Ibid.
97 Ibid.
98 www.usmagazine.com/entertainment/news/rep-russell-crowe-is-not-dead-2010106
99 http://abcnews.go.com/Entertainment/story?id=102477&page=1#.TzdQRm0Zmx8 http://entertainment.msnbc.msn.com/_news/2012/03/14/10685342-why-no-word-yet-on-whitney-houstons-cause-of-death
100 www.snopes.com/inboxer/hoaxes/bieber.asp
101 *Picasso (Masterpieces: Artists and Their Works)* by Shelley Swanson Sateren (Capstone Press, 2006)
102 http://socalhistory.org/biographies/howard-hughes.html
103 http://mentalfloss.com/amazingfactgenerator/
104 Ibid.
105 Ibid.
106 Ibid.
107 "The Prologue to Assassination: Rare Photograph Shows Plotters Present at Lincoln Inauguration," *Life*, February 13, 1956
108 www.msnbc.msn.com/id/6875436/#.T2JzpSO3CB9
109 *That's Not in My American History Book: A Compilation of Little Known Events and Forgotten Heroes* by Thomas Ayres (Taylor Trade Publications, 2004)
110 Ibid.
111 Ibid.
112 http://nextround.net/2009/03/chuck-norris-taught-bob-barker-karate
113 www.miqel.com/entheogens/francis_crick_dna_lsd.html
114 www.imdb.com/name/nm0000115/bio
115 Ibid.
116 *That's Not in My American History Book: A Compilation of Little*

Known Events and Forgotten Heroes by Thomas Ayres (Taylor Trade Publications, 2004)
117 Ibid.
118 www.imdb.com/name/nm0949574/bio
119 http://listverse.com/2011/02/14/top-10-celebrities-who-have-killed-someone/
120 Ibid.
121 Ibid.
122 Ibid.
123 Ibid.
124 Ibid.
125 www.criminaljusticeschools.com/blog/21-celebrities-who-have-killed
126 Ibid.
127 Ibid.
128 Ibid.
129 www.reuters.com/article/2008/02/10/music-dion-dc-idUSN0948472220080210?pageNumber=1
130 http://news.bbc.co.uk/2/hi/entertainment/6278145.stm
131 www.reuters.com/article/2008/02/10/music-dion-dc-idUSN0948472220080210?pageNumber=1
132 www.allbusiness.com/retail-trade/miscellaneous-retail-retail-stores-not/4599216-1.html
133 www.fashionwindows.com/beauty/2003B/celine_dion02.asp
134 www.stylelist.com/2010/04/02/celine-dion-new-perfume-pure-brilliance/
135 www.celinedion.com
136 http://jam.canoe.ca/Music/Pop_Encyclopedia/D/Dion_Celine.html
137 Ibid.
138 www.encyclopedia.com/topic/Celine_Dion.aspx
139 Ibid.
140 Ibid.
141 celinedion.com
142 *Let's Talk About Love: A Journey to the End of Taste* by Carl Wilson (Continuum, 2007)
143 Ibid.
144 www.washingtonpost.com/blogs/celebritology/post/celine-dions-lawyer-ends-ridiculous-pictures-tumblr/2011/07/22/gIQAy4O0TI_blog.html
145 www.criminaljusticeschools.com/blog/21-celebrities-who-have-killed
146 Ibid.
147 Ibid.
148 www.guardian.co.uk/media/2006/dec/09/broadcasting.frontpagenews
149 www.criminaljusticeschools.com/blog/21-celebrities-who-have-killed
150 Ibid.

151 Ibid.
152 Ibid.
153 Ibid.
154 *That's Not in My American History Book: A Compilation of Little Known Events and Forgotten Heroes* by Thomas Ayres (Taylor Trade Publications, 2004)
155 Ibid.
156 Ibid.
157 *The Pocket Idiot's Guide to Not So Useless Facts* by Dane Sherwood, Sandy Wood and Kara Kovalchik (Alpha Books, 2006)
158 Ibid.

THE 99 PERCENT

159 *DWI, DUI and the Law* by Margaret C. Jasper (Oceana Publications, 2004)
160 www.msnbc.msn.com/id/32321637/ns/health-addictions/t/women-drinking-more-duis-experts-say/
161 Ibid.
162 http://abcnews.go.com/GMA/story?id=8350259
163 www.rcpsych.ac.uk/pdf/MenBehavingSadly.pdf
164 www.scientificamerican.com/podcast/episode.cfm?id=women-apologize-more-frequently-tha-10-09-25
165 www.pcworld.com/article/144129/internet_fraud_dupes_men_more_often_than_women.html
166 www.cbsnews.com/stories/2010/05/19/health/main6499561.shtml
167 www.popsci.com/scitech/article/2009-09/are-men-or-women-more-likely-be-hit-lightning
168 http://onlinelibrary.wiley.com/doi/10.1111/1467-9515.00241/abstract
169 www.afsp.org/index.cfm?page_id=04ECB949-C3D9-5FFA-DA9C65C381BAAEC0
170 Ibid.
171 Ibid.
172 Ibid.
173 Ibid.
174 www.cosmopolitan.com/advice/tips/great-female-survey/satisfying-sex-lives
175 Ibid.
176 www.who.int/mental_health/prevention/genderwomen/en/
177 Ibid.
178 Ibid.
179 Ibid.
180 www.cmu.edu/homepage/collaboration/2007/winter/traffic-stats.shtml

181 United Nations Development Programme, "Human Development Report," 2007
182 *The Gendered Nature of Natural Disasters: The Impact of Catastrophic Events on the Gender Gap in Life Expectancy, 1981–2002* by E. Neumayer and T. Plumper (London School of Economics and Political Science with University of Essex and Max-Planck Institute of Economics, 2007); "Reaching Out to Women When Disaster Strikes" by K. Peterson (Soroptimist White Paper, 2007)
183 Ibid.
184 http://familysafemedia.com/pornography_statistics.html#anchor4
185 www.hsph.harvard.edu/news/press-releases/archives/2003-releases/press06062003.html
186 www.enotes.com/nursing-encyclopedia/blood
187 www.darwinawards.com
188 Ibid.
189 Ibid.
190 Ibid.
191 *Discover's 20 Things You Didn't Know About Everything: Duct Tape, Airport Security, Your Body, Sex in Space . . . and More!* by the editors of *Discover* magazine and Dean Christopher (HarperCollins, 2008)
192 Ibid.
193 Ibid.
194 www.neatorama.com/2007/03/12/30-strangest-deaths-in-history
195 Ibid.
196 Ibid.
197 Ibid.
198 www.nytimes.com/1985/08/02/us/victim-at-lifeguards-party.html
199 www.pet-abuse.com/cases/5034
200 www.ctv.ca/CTVNews/TopStories/20100701/brazil-sex-car-100701/
201 www.usatoday.com/news/nation/2007-06-20-naked-deaths_N.htm
202 http://adventure.howstuffworks.com/survival/wilderness/niagara6.htm
203 *Discover's 20 Things You Didn't Know About Everything: Duct Tape, Airport Security, Your Body, Sex in Space . . . and More!* by the editors of *Discover* magazine and Dean Christopher (HarperCollins, 2008)
204 www.snopes.com/horrors/freakish/onstage.asp
205 http://journalstar.com/news/local/article_d61cc109-3492-54ef-849d-0a5d7f48027a.html
206 www.dailymail.co.uk/news/article-442554/Hospital-patient-died-setting-medication-cigarette.html
207 http://dictionary.reference.com/browse/coulrophobia?s=t
208 www.psychologytoday.com/articles/200610/no-laughing-matter
209 www.wisegeek.com/how-common-is-coulrophobia.htm
210 http://news.bbc.co.uk/2/hi/uk_news/magazine/7191721.stm
211 www.wisegeek.com/what-is-coulrophobia.htm

212 http://news.bbc.co.uk/2/hi/uk_news/magazine/7191721.stm
213 www.wisegeek.com/what-is-coulrophobia.htm
214 http://news.bbc.co.uk/2/hi/uk_news/magazine/7191721.stm
215 *Christmas Sucks: What to Do When Fruitcake, Family and Finding the Perfect Gift Make You Miserable* by Joanne Kimes (Adams, 2008)
216 www.biography.com/people/john-wayne-gacy-10367544
217 Ibid.
218 http://phobias.about.com/od/introductiontophobias/a/clownphobia.h
219 *An Excess of Phobias and Manias: A Compilation of Anxieties, Obsessions and Compulsions That Push Many Over the Edge of Sanity* by John G. Robertson (Senior Scribe Publications, 2003)
220 www.economist.com/node/18713690
221 www.psychologytoday.com/articles/200610/no-laughing-matter
222 www.guardian.co.uk/society/joepublic/2010/feb/11/coulrophobia-clowns-sc; http://listverse.com/2010/11/03/10-well-known-people-and-their-phobias; http://abcnews.go.com/Entertainment/WolfFiles/story?id=116591&page=1
223 www.guardian.co.uk/society/joepublic/2010/feb/11/coulrophobia-clowns-sc
224 Ibid.
225 http://knowyourmeme.com/memes/lying-down-game#.TmFilpiGOkB
226 Ibid.
227 Ibid.
228 www.washingtonpost.com/blogs/blogpost/post/is-planking-connected-to-the-slave-trade/2011/07/08/gIQAz1aj3H_blog.html
229 Ibid.
230 http://knowyourmeme.com/memes/planking#.TmFlcJiGOkA
231 Ibid.
232 www.twitter.com/#!/AnnaSophiaB/status/98597541880406016
233 www.timesonline.co.uk/tol/life_and_style/health/article6827618.ece
234 www.washingtonpost.com/blogs/blogpost/post/is-planking-connected-to-the-slave-trade/2011/07/08/gIQAz1aj3H_blog.html
235 Ibid.
236 www.dailymail.co.uk/news/article-1387272/Planking-claims-victim-Acton-Beale-falls-balcony-death.html
237 Ibid.
238 Ibid.
239 http://tvnz.co.nz/national-news/plank-key-hit-facebook-4199767
240 http://knowyourmeme.com/memes/planking#.TmFlcJiGOkA
241 www.examiner.com.au/news/national/national/general/forget-planking-now-its-teapotting/2168552.aspx; http://edition.cnn.com/2011/TECH/web/07/21/owling.meme/index.html?hpt=hp_t2

242 www.askmen.com/top_10/entertainment/187_top_10_list.html
243 *Tennessee: A Guide to the State* by the Federal Writers' Project (Viking Press, 1939)
244 Ibid.
245 Ibid.
246 www.popcrunch.com/14-freakiest-serial-killers-youve-never-heard-of/
247 http://listverse.com/2011/02/09/10-evil-psychopaths-you-probably-dont-know/
248 Ibid.
249 Ibid.
250 www.askmen.com/top_10/entertainment/187_top_10_list.html
251 http://listverse.com/2011/02/09/10-evil-psychopaths-you-probably-dont-know/
252 Ibid.
253 www.askmen.com/top_10/entertainment/187_top_10_list.html
254 Ibid.
255 http://listverse.com/2011/02/09/10-evil-psychopaths-you-probably-dont-know/
256 www.askmen.com/top_10/entertainment/187_top_10_list.html
257 http://listverse.com/2011/02/09/10-evil-psychopaths-you-probably-dont-know/
258 www.popcrunch.com/14-freakiest-serial-killers-youve-never-heard-of/
259 www.askmen.com/top_10/entertainment/187_top_10_list.html
260 Ibid.
261 Ibid.
262 http://listverse.com/2011/02/09/10-evil-psychopaths-you-probably-dont-know/; www.trutv.com/library/crime/serial_killers/history/earle_nelson/1.html
263 Ibid.
264 Ibid.
265 Ibid.
266 www.popcrunch.com/14-freakiest-serial-killers-youve-never-heard-of/
267 Ibid.
268 "World's Worst Killers," BBC News, http://news.bbc.co.uk/2/hi/495477.stm
269 Ibid.
270 www.popcrunch.com/14-freakiest-serial-killers-youve-never-heard-of/
271 World's Worst Killers. BBC News. http://news.bbc.co.uk/2/hi/495477.stm
272 Ibid.
273 www.popcrunch.com/14-freakiest-serial-killers-youve-never-heard-of/

274 Ibid.
275 "World's Worst Killers," BBC News, http://news.bbc.co.uk/2/hi/495477.stm
276 Ibid.; http://crime.about.com/od/serial/p/lobez.htm
277 Ibid.
278 "World's Worst Killers," BBC News, http://news.bbc.co.uk/2/hi/495477.stm
279 www.smh.com.au/news/film/up-the-creek/2005/10/27/1130367983912.html
280 "World's Worst Killers," BBC News, http://news.bbc.co.uk/2/hi/495477.stm
281 "Profiler," interview of John E. Douglas by David Bowman, Salon.com, July 8, 1999; "Buffalo Bill" by Anthony Bruno, *All About Hannibal Lecter: Facts and Fiction*, page 2, CrimeLibrary.com
282 www.popcrunch.com/14-freakiest-serial-killers-youve-never-heard-of/
283 Ibid.
284 *The A to Z Encyclopedia of Serial Killers* by Harold Schechter and David Everitt (Simon and Schuster, 2006)
285 Ibid.; www.askmen.com/top_10/entertainment/187_top_10_list.html; www.mayhem.net/Crime/serial1.html
286 http://crime.about.com/od/serial/p/lobez.htm
287 Ibid.; *The Serial Killers: A Study in the Psychology of Violence* by Colin Wilson and Donald Seaman (Ebury Publishing, 2011)
288 http://crime.about.com/od/serial/p/albertfish.htm
289 http://icsahome.com/infoserv_respond/by_article.asp?ID=45015
290 www.msnbc.msn.com/id/27531105/ns/world_news-americas/t/dominican-migrant-we-ate-flesh-survive/
291 www.mayhem.net/Crime/cannibals1.html
292 Ibid.
293 www.spiegel.de/international/zeitgeist/0,1518,511775,00.html; http://news.bbc.co.uk/2/hi/europe/3443803.stm
294 http://digitaljournal.com/article/258197
295 www.smithsonianmag.com/travel/cannibals.html
296 http://works.bepress.com/carmen_cusack/4/
297 *1607: Jamestown and the New World* by Dennis Montgomery (Colonial Williamsburg, 2007)
298 *In the Heart of the Sea: The Tragedy of the Whaleship Essex* by Nathaniel Philbrick (Penguin, 2001); *The Wreck of the Whaleship Essex* by Owen Chase (Pimlico, 2000)
299 www.news.com.au/entertainment/movies/film-about-tasmanian-cannibal-alexander-pearce-makes-audience-vomit/story-e6frfmvr-1225775826175
300 *Hungry Ghosts: Mao's Secret Famine* by Jasper Becker (Macmillan, 1998); www.history.com/this-day-in-history/siege-of-leningrad-begins

301 “Cannibalism—Without a Hangman, Without a Rope: Navy War Crimes Trials After World War II” by J. M. Welch, *International Journal of Naval History* 1, no. 1 (April 2002)
302 *Flyboys: A True Story of Courage* by James Bradley (Back Bay Books, 2004), pp. 229–30, 311, 404
303 www.viven.com.uy/571/eng/accidente.asp
304 www.pbs.org/wgbh/americanexperience/features/transcript/donner-transcript
305 www.consumethisfirst.com/2010/06/10/worst-food-of-the-week-country-time-lemonade/
306 *Bigger Secrets: More Than 125 Things They Prayed You'd Never Find Out* by William Poundstone (Houghton Mifflin Co, 1989)
307 http://healthland.time.com/2011/10/27/why-lovin-the-mcrib-isnt-a-heart-smart-idea/
308 http://nutrition.mcdonalds.com/nutritionexchange/nutritionfacts.pdf
309 www.icanhasinternets.com/2010/05/20-awesome-bacon-facts-graphic
310 Ibid.
311 www.fsis.usda.gov/Help/glossary-B/index.asp
312 www.salon.com/life/feature/2008/07/07/bacon_mania/
313 www.webmd.com/food-recipes/features/can-bacon-be-part-of-a-healthy-diet
314 Ibid.
315 Ibid.
316 Ibid.
317 Ibid.
318 www.webmd.com/news/20070417/study-copd-cured-meats-may-be-linked
319 www.lungusa.org/lung-disease/copd/resources/facts-figures/COPD-Fact-Sheet.html
320 www.webmd.com/food-recipes/features/can-bacon-be-part-of-a-healthy-diet
321 http://circ.ahajournals.org/cgi/content/abstract/CIRCULATIONAHA.109.924977v1?maxtoshow=&hits=10&RESULTFORMAT=&fulltext=processed+meat&searchid=1&FIRSTINDEX=0&resourcetype=HWCIT
322 www.cnn.com/2010/HEALTH/03/28/fatty.foods.brain/index.html
323 www.fbi.gov/news/stories/2011/may/predators_051711/predators_051711
324 Ibid.
325 www.fbi.gov/stats-services/publications/innocent-images-1
326 www.bullyingstatistics.org/content/cyber-bullying-statistics.html
327 Ibid.
328 Ibid.
329 www.privacyrights.org/fs/fs14-stk.htm

330 www.spendonlife.com/guide/identity-theft-statistics
331 Ibid.
332 Ibid.
333 Ibid.
334 Ibid.
335 Ibid.
336 www.coldcreekwellness.com/addiction-recovery/facts-about-internet-addiction-across-the-world/
337 Ibid.
338 Ibid.
339 Ibid.
340 Ibid.
341 Ibid.
342 www.pcsndreams.com/Pages/Virus_protection.htm
343 http://enough.org/inside.php?id=3K03RC4L00#2
344 Ibid.
345 Ibid.
346 Ibid.
347 www.cfr.org/terrorism-and-technology/terrorists-internet/p10005
348 Ibid.
349 Twitter.com
350 www.time.com/time/business/article/0,8599,1998055,00.html#ixzz1cVqzGAQx
351 Ibid.
352 "The Nocebo Effect on the Web: An Analysis of Fake Anti-Virus Distribution" by Moheeb Abu Rajab and Luca Ballard (Google, April 13, 2010)
353 http://1000memories.com/blog/94-number-of-photos-ever-taken-digital-and-analog-in-shoebox
354 www.bbc.co.uk/news/uk-wales-12013756
355 *Firestorm at Peshtigo: A Town, Its People and the Deadliest Fire in American History* by Denise Gess and William Lutz (Macmillan, 2003)
356 www.cbsnews.com/8301-504083_162-20055660-504083.html?tag=mncol;lst;2
357 http://news.bbc.co.uk/2/hi/uk_news/politics/4204980.stm
358 Ibid.
359 www.schillerinstitute.org/programs/program_symp_2_7_98_tchor_.html
360 http://articles.sfgate.com/2005-01-23/bay-area/17357897_1_prisons-inmate-cell-doors
361 Ibid.
362 www.lifegem.com
363 www.heavenlystarsfireworks.com; www.news-journalonline.com/news/local/flagler/2011/10/15/flagler-fireworks-firm-promotes-new-trend-in-final-sendoff.html

364 http://bls.gov/news.release/cfoi.nr0.htm
365 Ibid.
366 Ibid.
367 Ibid.
368 *The Pocket Idiot's Guide to Not So Useless Facts* by Dane Sherwood, Sandy Wood and Kara Kovalchik (Alpha Books, 2006)

OH, THE PLACES YOU'LL GO...IN YOUR PANTS

369 http://travel.yahoo.com/ideas/worlds-creepiest-places-045945458.html
370 Ibid.
371 http://list25.com/the-25-most-dangerous-cities-on-earth/5/
372 Ibid.
373 Ibid.
374 Ibid.
375 Ibid.
376 Ibid.
377 http://alaska.gov/kids/learn/facts.htm
378 www.ebizarre.com/Category/Animals_and_Creatures/11/
379 www.americashealthrankings.org
380 http://alaska.gov/kids/learn/facts.htm
381 http://commerce.alaska.gov/ded/dev/student_info/learn/facts.htm
382 Ibid.
383 http://hss.state.ak.us/dph/bvs/death_statistics/Leading_Causes_Census/frame.html
384 www.alaskacitizensforjustice.com/current-crime-lab.php
385 www.welcometoalaska.com/facts.htm
386 Ibid.
387 www.50states.com/facts/alaska.htm
388 www.welcometoalaska.com/facts.htm
389 www.epa.gov/oem/content/learning/exxon.htm
390 www.defenders.org/wildlife_and_habitat/wildlife/grizzly_bear.php
391 www.vpc.org
392 Ibid.
393 www.americashealthrankings.org
394 http://earthquake.usgs.gov/earthquakes/states/events/1964_03_28.php
395 http://wcatwc.arh.noaa.gov/64quake.htm
396 http://earthquake.usgs.gov/earthquakes/states/events/1964_03_28.php
397 Ibid.; http://wcatwc.arh.noaa.gov/64quake.htm
398 http://earthquake.usgs.gov/earthquakes/states/events/1964_03_28.php
399 Ibid.

400 http://articles.nydailynews.com/2010-10-28/entertainment/29441285_1_halloween-party-skeletons-bones
401 Ibid.
402 Ibid.
403 Ibid.
404 Ibid.
405 www.ouramazingplanet.com/1567-7-most-dangerous-places-earth-natural-disasters.html
406 Ibid.
407 Ibid.
408 Ibid.
409 www.ouramazingplanet.com/1567-7-most-dangerous-places-earth-natural-disasters.html
410 www.mineofuseless.info/trivia/Geography/?p=7
411 Ibid.
412 http://thebudingroup.com/new/budingroup/content.asp?contentid=2017298946
413 *Social Welfare: A History of the American Response to Need* by June Axinn and Mark J. Stern (Pearson/Allyn and Bacon, 2007)
414 www.factsurf.com/lists/details/7/10-Quick-Facts-About-New-York
415 Ibid.
416 www.crazy-laws.com/New-York%20_crazy_laws.htm
417 Ibid.
418 Ibid.
419 Ibid.
420 Ibid.
421 Ibid.
422 www.askmen.com/entertainment/special_feature_250/275_5-things-you-didnt-know-new-york-city.html
423 www.nyc.gov/html/dcp/pdf/census/nny_briefing_booklet.pdf
424 *Strange But True (New York City: Tales of the Big Apple)* by S. B. Howard (Globe Pequot, 2005)
425 www.askmen.com/entertainment/special_feature_250/275_5-things-you-didnt-know-new-york-city.html
426 *Strange But True (New York City: Tales of the Big Apple)* by S. B. Howard (Globe Pequot, 2005)
427 Ibid.
428 Ibid.
429 Ibid.
430 Ibid.
431 Ibid.
432 Ibid.
433 Ibid.
434 Ibid.
435 Ibid.

436 www.nyc.gov/html/oem/html/hazards/weather_thunder.shtml; http://abcnews.go.com/Technology/lightning-strikes-empire-state-building-times-row-video/story?id=13374451
437 http://mta.info/nyct/facts/ridership/index.htm
438 www.gothamgazette.com/article/issueoftheweek/20071126/200/2356
439 *Strange But True (New York City: Tales of the Big Apple)* by S. B. Howard (Globe Pequot, 2005)
440 www.environmentalgraffiti.com/travel/news-five-most-lawless-lands-earth
441 Ibid.
442 Ibid.; www.state.gov/j/drl/rls/hrrpt/2010/wha/154499.htm
443 www.environmentalgraffiti.com/travel/news-five-most-lawless-lands-earth; http://insightcrime.org/insight-latest-news/item/395-colombia-by-the-numbers-2010
444 www.environmentalgraffiti.com/travel/news-five-most-lawless-lands-earth
445 Ibid.
446 Ibid.
447 www.travelandleisure.com/articles/the-worlds-most-dangerous-countries
448 Ibid.
449 Ibid.
450 Ibid.
451 Ibid.
452 www.forbes.com/2010/01/14/most-dangerous-countries-lifestyle-travel-haiti-afghanistan-iraq.html
453 Ibid.
454 www.ouramazingplanet.com/1567-7-most-dangerous-places-earth-natural-disasters.html
455 http://articles.nydailynews.com/2010-10-28/entertainment/29441285_1_halloween-party-skeletons-bones
456 Ibid.
457 Ibid.
458 www.environmentalgraffiti.com/travel/news-five-most-lawless-lands-earth
459 Ibid.
460 http://travel.yahoo.com/ideas/worlds-creepiest-places-045945458.html
461 Ibid.
462 www.50states.com/facts/alaska.htm
463 www.texastribune.org/texas-education/public-education/texans-dinosaurs-humans-walked-the-earth-at-same/
464 Ibid.
465 Ibid.

466 http://jonathanturley.org/2009/09/08/constitutional-illiteracy-texas-orders-all-schools-to-teach-bible-literacy/
467 www.rasmussenreports.com/public_content/politics/general_state_surveys/texas/in_texas_31_say_state_has_right_to_secede_from_u_s_but_75_opt_to_stay
468 www.rasmussenreports.com/public_content/politics/general_state_surveys/texas/70_in_texas_favor_offshore_oil_drilling_66_support_deepwater_drilling
469 www.rasmussenreports.com/public_content/politics/general_state_surveys/texas/69_in_texas_support_using_u_s_troops_on_border
470 www.rasmussenreports.com/public_content/politics/elections/election_2010/election_2010_governor_elections/texas/56_in_texas_favor_state_lawsuit_to_stop_health_care_plan
471 *Fort Worth Star-Telegram*, June 14, 2007, via DeathPenaltyInfo.org
472 *Dallas Morning News*, March 16, 2003, via DeathPenaltyInfo.org
473 *Houston Chronicle*, December 31, 2002, via DeathPenaltyInfo.org
474 www.freerepublic.com/focus/f-news/976646/posts
475 Ibid.
476 www.texastribune.org/texas-taxes/2011-budget-shortfall/uttexas-tribune-poll-mixed-signals-on-budget-cuts/
477 www.dfwairport.com/contact/
478 Ibid.
479 www.legendsofamerica.com/tx-facts.html
480 Ibid.
481 http://gorvtexas.com/texasstrange.htm
482 Ibid.
483 *The Book of the Bizarre: Freaky Facts and Strange Stories* by Varla Ventura (Weiser, 2008)
484 "'Cheerleader Mom' Freed After Serving Six Months," by Michelle Koiden, *Abilene Reporter-News*, March 1, 1997
485 Ibid.
486 Google Maps, http://bit.ly/IGAhkQ
487 www.sciencedaily.com/releases/2010/09/100928122604.htm
488 *Amazing Texas* by T. Jensen Lacey (Jefferson Press, 2008)
489 Ibid.
490 http://bls.gov/news.release/cfoi.nr0.htm
491 www.ouramazingplanet.com/1567-7-most-dangerous-places-earth-natural-disasters.html
492 Ibid.
493 Ibid.
494 Ibid.
495 Ibid.
496 Ibid.
497 www.concierge.com/ideas/holidays/tours/1563
498 Ibid.

499 Ibid.
500 www.gadling.com/2011/04/27/the-worlds-ten-creepiest-abandoned-cities/
501 Ibid.
502 Ibid.
503 http://travel.yahoo.com/ideas/worlds-creepiest-places-045945458.html
504 http://urbantitan.com/10-most-dangerous-cities-in-the-world-in-2011/
505 Ibid.
506 Ibid.
507 Ibid.
508 Ibid.
509 Ibid.
510 Ibid.

THE BETTER TO EAT YOU WITH, MY DEAR

511 http://animals.nationalgeographic.com/animals/reptiles/nile-crocodile/
512 Ibid.
513 http://news.bbc.co.uk/onthisday/hi/dates/stories/october/29/newsid_2467000/2467665.stm
514 www.independent.co.uk/news/world/australasia/fear-of-the-dingo-returns-to-australia-in-wake-of-boys-death-683861.html
515 http://animals.howstuffworks.com/animal-facts/piranha-eat-cows1.htm
516 www.dailymail.co.uk/news/article-2042152/Piranhas-attack-beach-Brazil-100-swimmers-bitten-1-weekend.html?ITO=1490
517 http://animals.howstuffworks.com/animal-facts/piranha-eat-cows1.htm
518 www.fws.gov/endangered/esa-library/pdf/polar_bear.pdf
519 http://animals.nationalgeographic.com/animals/mammals/grizzly-bear/
520 *Bear Attacks: Their Causes and Avoidance* by Stephen Herrero, revised edition (McClelland & Stewart, 2003)
521 Ibid.
522 *The Living Animals of the World: A Popular Natural History with One Thousand Illustrations, vol. 1: Mammals* by C. J. Cornish, Frederick Courteney Selous, Sir Harry Hamilton Johnston and Sir Herbert Maxwell (Dodd, Mead and Company), Archive.org
523 http://articles.cnn.com/2011-07-07/us/wyoming.grizzly.attack_1_bear-attacks-kerry-gunther-yellowstone-ecosystem?_s=PM:US
524 www.cbsnews.com/stories/2003/10/08/national/main577043.shtml

525 http://animals.nationalgeographic.com/animals/fish/great-white-shark/
526 Ibid.
527 www.flmnh.ufl.edu/fish/sharks/statistics/gattack/mapusa.htm
528 www.taronga.org.au/animals-conservation/conservation-science/australian-shark-attack-file/latest-figures
529 www.flmnh.ufl.edu/fish/sharks/statistics/gattack/world.htm
530 www.straightdope.com/columns/read/2952/did-the-australian-army-once-wage-war-on-emus
531 www.msnbc.msn.com/id/27428859/ns/technology_and_science-science/t/ten-scariest-animals-nature/
532 Ibid.
533 Ibid.
534 Ibid.
535 http://bbcearth.posterous.com/top-ten-scariest-animals-plants
536 Ibid.
537 www.livescience.com/11325-top-10-deadliest-animals.html
538 www.squidoo.com/the-worlds-deadliest-animals#module56576142
539 Ibid.
540 www.huffingtonpost.com/2009/10/17/the-deadliest-animalswhic_n_322321.html
541 www.straightdope.com/columns/read/1862/are-hippos-the-most-dangerous-animal; www.sandiegozoo.org/animalbytes/t-hippopotamus.html
542 www.straightdope.com/columns/read/1862/are-hippos-the-most-dangerous-animal
543 www.sandiegozoo.org/animalbytes/t-hippopotamus.html
544 www.straightdope.com/columns/read/1862/are-hippos-the-most-dangerous-animal
545 "The Dangerous Hippo," by George W. Frame and Lory Herbison Frame, *Science Digest* LXXVI (November 1974), pp. 80–86
546 Ibid.
547 *The New Book of Lists: The Original Compendium of Curious Information* by David Wallechinsky and Amy Wallace (Canongate U.S., 2005)
548 Ibid.
549 Ibid.
550 Ibid.
551 www.latimes.com/business/technology/la-fi-tn-squid-rocket-science-20120221,0,5248637.story
552 Ibid.
553 *The New Book of Lists: The Original Compendium of Curious Information* by David Wallechinsky and Amy Wallace (Canongate U.S., 2005)
554 Ibid.
555 Ibid.
556 *The Complete Cat's Meow: Everything You Need to Know About Caring for Your Cat* by Darlene Arden (John Wiley & Sons, 2011)

557 http://animals.howstuffworks.com/pets/domestic-cat-info9.htm
558 *Doc Halligan's What Every Pet Owner Should Know: Prescriptions for Happy, Healthy Cats and Dogs* by Karen Halligan (HarperCollins, 2008)
559 http://cats.about.com/cs/catmanagement101/a/how_cat_work_3.htm
560 www.animalpeoplenews.org/03/9/dogs.catseatenAsia903.html
561 *Encyclopedia of Cats* by Candida Frith-Macdonald (Parragon Books, 2008)
562 Ibid.
563 Ibid.
564 Ibid.
565 *Is My Cat a Tiger?* by Jenni Bidner (Lark Books, 2006)
566 *Encyclopedia of Cats* by Candida Frith-Macdonald (Parragon Books, 2008)
567 *Is My Cat a Tiger?* by Jenni Bidner (Lark Books, 2006)
568 www.cat-world.com.au/cat-world-records
569 Ibid.
570 Ibid.
571 *What Cats Are Made Of* by Hanoch Piven (Ginee Seo Books, 2009)
572 Ibid.
573 Ibid.
574 Ibid.
575 *Cat* by Juliet Clutton-Brock (DK Publishing, 2004)
576 Ibid.
577 Ibid.
578 *The Welfare of Cats* by Irene Rochlitz (Springer, 2005)
579 Ibid.
580 *The New Book of Lists: The Original Compendium of Curious Information* by David Wallechinsky and Amy Wallace (Canongate U.S., 2005)
581 Ibid.
582 Ibid.
583 Ibid.
584 Ibid.
585 Ibid.
586 Ibid.
587 http://listverse.com/2007/12/16/top-10-animals-you-didnt-know-were-venomous/
588 Ibid.
589 Ibid.
590 Ibid.
591 http://coedmagazine.com/2011/04/06/6-animals-you-didnt-know-were-dangerous
592 www.womansday.com/life/10-surprisingly-lethal-animals-10820
593 Ibid.
594 Ibid.

595 Ibid.
596 Ibid.
597 Ibid.
598 Ibid.
599 www.msnbc.msn.com/id/14663786/ns/world_news-asia_pacific/t/crocodile-hunter-steve-irwin-killed-stingray/
600 www.omg-facts.com/view/Facts/22383
601 www.wildwatch.com/living_library/mammals-2/honey-badger
602 http://animaldiversity.ummz.umich.edu/site/accounts/information/Mellivora_capensis.html
603 Ibid.; www.badgers.org.uk/badgerpages/honey-badger.html
604 http://animaldiversity.ummz.umich.edu/site/accounts/information/Mellivora_capensis.html
605 Ibid.
606 http://ngm.nationalgeographic.com/ngm/0409/feature6/index.html
607 www.badgers.org.uk/badgerpages/honey-badger.html; *East African Mammals: An Atlas of Evolution in Africa*, vol. 3, part 1, by Jonathan Kingdon (University of Chicago Press, 1988)
608 www.badgers.org.uk/badgerpages/honey-badger.html; *The Great Adventure: The University of California Southern Africa Expedition of 1947–1948* by Thomas John Larson (iUniverse, 2004)
609 www.badgers.org.uk/badgerpages/honey-badger.html
610 *The Great Adventure: The University of California Southern Africa Expedition of 1947–1948* by Thomas John Larson (iUniverse, 2004)
611 http://ngm.nationalgeographic.com/ngm/0409/feature6/index.html
612 Ibid.
613 www.badgers.org.uk/badgerpages/honey-badger.html
614 http://animaldiversity.ummz.umich.edu/site/accounts/information/Mellivora_capensis.html
615 www.awf.org/content/wildlife/detail/ratel; www.eol.org/pages/328035/details#behavior
616 http://ngm.nationalgeographic.com/ngm/0409/feature6/index.html
617 Ibid.
618 Ibid.
619 Ibid.
620 *East African Mammals: An Atlas of Evolution in Africa*, vol. 3, part 1, by Jonathan Kingdon (University of Chicago Press, 1988)
621 www.womansday.com/life/10-surprisingly-lethal-animals-10820
622 Ibid.
623 Ibid.
624 Ibid.
625 Ibid.
626 "Stitching Wounds with the Mandibles of Ants and Beetles" by E. W. Gudger, *Journal of the American Medical Association* 84 (1861–64)
627 www.zmescience.com/research/how-scientists-tught-monkeys-the

-concept-of-money-not-long-after-the-first-prostitute-monkey-appeared/
628 Ibid.
629 http://animal.discovery.com/news/briefs/20050711/chicken.html
630 Ibid.
631 *War Elephants* by J. Kistler (University of Nebraska Press, 2005, 2007); *Greek Fire, Poison Arrows and Scorpion Bombs: Biological and Chemical Warfare in the Ancient World* by A. Mayor (Overlook/ Duckworth, 2005, 2009)
632 www.csa.com/discoveryguides/snakehead/overview.php
633 www.msnbc.msn.com/id/27428859/ns/technology_and_science-science/t/ten-scariest-animals-nature/
634 http://neuro.bcm.edu/eagleman/asp/
635 www.msnbc.msn.com/id/27428859/ns/technology_and_science-science/t/ten-scariest-animals-nature
636 http://coedmagazine.com/2011/04/06/6-animals-you-didnt-know-were-dangerous
637 Ibid.
638 www.mnn.com/earth-matters/animals/photos/15-cute-animals-that-could-kill-you
639 Ibid.
640 Ibid.
641 Ibid.
642 www.treehugger.com/natural-sciences/5-endangered-species-that-could-kill-you-and-how-to-save-yourself.html
643 http://abcnews.go.com/GMA/AmazingAnimals/whale-kills-trainer-sea-worlds-shamu-stadium/story?id=9932526
644 Ibid.
645 *The Pocket Idiot's Guide to Not So Useless Facts* by Dane Sherwood, Sandy Wood and Kara Kovalchik (Alpha Books, 2006)
646 Ibid.

CORPUS HORRIFICUS

647 http://discovermagazine.com/2007/sep/20-things-you-didnt-know-about-hygiene
648 Ibid.
649 Ibid.
650 Ibid.
651 Ibid.
652 Ibid.
653 www.cdc.gov/healthywater/hygiene/fast_facts.html
654 http://discovermagazine.com/2007/sep/20-things-you-didnt-know-about-hygiene
655 Ibid.

656 Ibid.
657 Ibid.
658 Ibid.
659 Ibid.
660 www.usatoday.com/news/health/2009-03-10-dental-skip_N.htm
661 Ibid.
662 Ibid.
663 www.webmd.com/oral-health/guide/bad-breath; http://bodyandhealth.canada.com/channel_condition_info_details.asp?disease_id=66&channel_id=9&relation_id=10860
664 www.webmd.com/oral-health/guide/bad-breath; http://bodyandhealth.canada.com/channel_condition_info_details.asp?disease_id=66&channel_id=9&relation_id=10860
665 www.webmd.com/oral-health/guide/bad-breath
666 www.dailymail.co.uk/femail/article-335403/Is-chocolate-bad-breath.html
667 Ibid.
668 www.dentistparma.com/blog/post/little-known-facts-about-bad-breath.html
669 www.webmd.com/oral-health/guide/bad-breath
670 Ibid.
671 Ibid.
672 www.dentistparma.com/blog/post/little-known-facts-about-bad-breath.html
673 www.webmd.com/oral-health/guide/bad-breath
674 Ibid.
675 www.dentistparma.com/blog/post/little-known-facts-about-bad-breath.html
676 www.foxnews.com/story/0,2933,479231,00.html
677 www.news-medical.net/news/2007/12/14/33551.aspx
678 www.mentalfloss.com/blogs/archives/20221
679 www.dailymail.co.uk/news/article-1234784/He-stand-pain-Nazi-records-Hitler-hated-going-dentist.html
680 www.jamaicaobserver.com/news/Bad-breath-beating-gives-woman-6000-fine
681 http://transcripts.cnn.com/TRANSCRIPTS/1006/05/lkl.01.html
682 http://discovermagazine.com/2007/sep/20-things-you-didnt-know-about-hygiene
683 Ibid.
684 Ibid.
685 www.cdc.gov/healthywater/hygiene/fast_facts.html
686 Ibid.
687 Ibid.
688 Ibid.
689 http://health.howstuffworks.com/wellness/men/hygiene/5-mens-hygiene-facts-you-wont-believe.htm

690 Ibid.
691 www.webmd.com/oral-health/guide/bad-breath
692 www.cdc.gov/parasites/lice/
693 Ibid.
694 www.economist.com/node/15060097?subjectid=7933596&story_id=15060097
695 www.medicinenet.com/script/main/art.asp?articlekey=144552
696 Ibid.
697 http://my.clevelandclinic.org/disorders/penile_disorders/hic_disorders_of_the_penis.aspx; www.stdservices.on.net/std/balanitis/facts.htm
698 http://my.clevelandclinic.org/disorders/penile_disorders/hic_disorders_of_the_penis.aspx
699 Ibid.
700 Ibid.
701 Ibid.
702 www.webmd.com/sexual-conditions/tc/bacterial-vaginosis-topic-overview
703 www.health.com/health/condition-article/0,,20189833,00.html
704 www.webmd.com/sexual-conditions/tc/bacterial-vaginosis-topic-overview
705 www.health.com/health/library/mdp/0,,stv8866,00.html
706 www.lovelyish.com/732384028/8-interesting-facts-about-your-period/
707 www.foxnews.com/health/2010/03/29/shocking-facts-flow-wanted-know-period-afraid-ask/
708 http://blogs.scientificamerican.com/bering-in-mind/2010/03/01/a-bushel-of-facts-about-the-uniqueness-of-human-pubic-hair/
709 www.msnbc.msn.com/id/31144530/ns/health-womens_health/#.TqRni3HW2jo; *Primary Care of Women: A Guide for Midwives and Women's Health Providers* by Barbara Hackley, Jan M. Kriebs and Mary Ellen Rousseau (Jones & Bartlett, 2006)
710 http://health.howstuffworks.com/skin-care/beauty/hair-removal/brazilian-wax3.htm
711 www.msnbc.msn.com/id/31144530/ns/health-womens_health/#.TqRni3HW2jo
712 "Do Bikini Waxes Spread STDs?" *Shape*, August 2009, http://findarticles.com/p/articles/mi_m0846/is_12_28/ai_n32429307/
713 www.cosmopolitan.co.uk/lifestyle/25-facts-about-boobs-100931
714 Ibid.
715 www.womens-health-issues.us/breast-facts.php
716 Ibid.
717 www.pixlmonster.com/tully/breasts/
718 Ibid.
719 www.cosmopolitan.co.uk/lifestyle/25-facts-about-boobs-100931

720 www.womens-health-issues.us/breast-facts.php
721 www.pixlmonster.com/tully/breasts/
722 www.msnbc.msn.com/id/21599854/#.TkCDiL-GOkA
723 www.cosmopolitan.co.uk/lifestyle/25-facts-about-boobs-100931
724 Ibid.
725 Ibid.
726 www.pixlmonster.com/tully/breasts/
727 Ibid.
728 Ibid.
729 Ibid.
730 Ibid.
731 Ibid.
732 www.foxnews.com/health/2010/07/14/woman-worlds-largest-breasts-fighting-life/
733 www.askmen.com/entertainment/special_feature_150/157b_special_feature.html
734 http://breastcancer.about.com/b/2008/02/05/saint-agatha-patron-saint-for-breast-cancer.htm; http://saints.sqpn.com/saint-agatha-of-sicily/
735 www.mayoclinic.com/health/morgellons-disease/sn00043
736 morgellons.org
737 http://rarediseases.info.nih.gov/GARD/Condition/8529/Macrodactyly_of_the_hand.aspx
738 www.ncbi.nlm.nih.gov/pubmedhealth/PMH0001411/
739 Ibid.
740 www.webmd.com/a-to-z-guides/jumping-frenchmen-of-maine
741 www.ncbi.nlm.nih.gov/pubmedhealth/PMH0002350; http://cdc.gov/dengue
742 www.cdc.gov/dengue
743 www.ncbi.nlm.nih.gov/pubmedhealth/PMH0002350
744 http://bmb.oxfordjournals.org/content/95/1/161; www.vagabondish.com/10-nastiest-travel-diseases/
745 www.smh.com.au/news/health-and-fitness/my-brother-is-an-alien/2006/02/08/1139379562020.html?page=fullpage
746 Ibid.
747 Ibid.
748 Ibid.
749 Ibid.
750 Ibid.
751 Ibid.
752 Ibid.
753 Ibid.
754 Ibid.
755 Ibid.
756 Ibid.; http://thehumanmarvels.com/?p=105

757 www.smh.com.au/news/health-and-fitness/my-brother-is-an-alien/2006/02/08/1139379562020.html?page=fullpage; http://emedicine.medscape.com/article/1072987-overview
758 www.smh.com.au/news/health-and-fitness/my-brother-is-an-alien/2006/02/08/1139379562020.html?page=fullpage
759 Ibid.
760 Ibid.
761 Ibid.
762 www.telegraph.co.uk/news/worldnews/europe/italy/7914746/Scientists-investigate-Stendhal-Syndrome-fainting-caused-by-great-art.html
763 www.moebiussyndrome.com
764 www.ncbi.nlm.nih.gov/pubmedhealth/PMH0002622
765 Ibid.
766 www.dailymail.co.uk/health/article-512416/The-girl-10-die-shock-just-watching-scary-film.html
767 http://emedicine.medscape.com/article/438994-overview
768 Ibid.
769 www.ncbi.nlm.nih.gov/pmc/articles/PMC1784613/
770 Ibid.
771 http://dermnetnz.org/immune/paraneoplastic-pemphigus.html
772 Ibid.
773 http://news.bbc.co.uk/2/hi/uk_news/wales/south_west/4335454.stm
774 www.ninds.nih.gov/disorders/prosopagnosia
775 www.ninds.nih.gov/disorders/prosopagnosia
776 www.ninds.nih.gov/disorders/kleine_levin/kleine_levin.htm
777 www.mdsupport.org
778 Ibid.
779 http://encyclo.co.uk/define/Diphallia
780 *Fregoli Delusion* by Frederic P Miller, Agnes F Vandome and John McBrewster, eds. (VDM Publishing, 2010)
781 www.msnbc.msn.com/id/15391010/ns/travel-news/t/paris-syndrome-leaves-tourists-shock/#.TpNqpHHW2jo
782 Ibid.
783 www.websters-online-dictionary.org/definitions/reduplicative?cx=partner-pub-0939450753529744%3Av0qd01-tdlq&cof=FORID%3A9&ie=UTF-8&q=reduplicative&sa=Search#906
784 Ibid.
785 www.portfolio.com/special-reports/2010/09/07/detroit-tops-list-of-most-stressful-metropolitan-areas/
786 http://discovermagazine.com/2007/sep/20-things-you-didnt-know-about-hygiene
787 Ibid.
788 Ibid.

789 Ibid.
790 Ibid.
791 Ibid.
792 Ibid.
793 *Traumatic Stress: The Effects of Overwhelming Experience on Mind, Body and Society* by Bessel A. Van der Kolk, et al. (Guilford Press, 2007)
794 Ibid.
795 Ibid.
796 *Mind, Stress and Emotion: The New Science of Mood* by Gene Wallenstein (Commonwealth Press, 2003)
797 Ibid.
798 *Heart Disease: An Essential Guide for the Newly Diagnosed* by Lawrence Chilnick (Perseus Books, 2008)
799 "Poll: Money Worries World's Greatest Cause of Stress," by Claire Barthelemy, CNN.com, September 30, 2009
800 www.fi.edu/learn/brain/index.html
801 http://news.bbc.co.uk/2/hi/4646010.stm
802 Ibid.
803 http://well.blogs.nytimes.com/2009/08/06/the-pain-of-being-a-redhead/
804 www.vivo.colostate.edu/hbooks/pathphys/digestion/stomach/vomiting.html
805 *The Pocket Guide to Girl Stuff* by Bart King (Gibbs Smith, 2009)
806 www.straightdope.com/columns/read/1108/is-uranium-added-to-false-teeth-to-give-them-a-natural-glow
807 www.surgery.org
808 www.medicalbillingandcoding.org/
809 www.surgery.org
810 Ibid.
811 Ibid.
812 www.medicalbillingandcoding.org/
813 Ibid.
814 Ibid.
815 Ibid.
816 Ibid.
817 Ibid.
818 Ibid.
819 www.forbes.com/2007/10/09/health-surgery-risks-forbeslife-cx_mlm_1010health.html
820 Ibid.
821 Ibid.
822 Ibid.
823 Ibid.
824 Ibid.
825 www.medicalbillingandcoding.org/

826 www.webmd.com/healthy-beauty/guide/cosmetic-procedures-botox
827 Ibid.
828 www.medicalbillingandcoding.org/
829 Ibid.
830 www.forbes.com/2007/10/09/health-surgery-risks-forbeslife-cx_mlm_1010health.html
831 Ibid.
832 www.sciencekids.co.nz/sciencefacts/humanbody.html
833 Ibid.
834 Ibid.

MAKING THE BEAST

835 http://emedicine.medscape.com/article/258768-overview
836 www.innerbody.com/image/repfov.html
837 *Sport, Culture and Advertising: Identities, Commodities and the Politics of Representation* by Steven J. Jackson and David L. Andrews (Psychology Press, 2005)
838 "Quadruplets and Higher Multiple Births," by Marie M. Clay, *Clinics in Developmental Medicine* 107 (Cambridge University Press, 1989)
839 http://women.webmd.com/picture-of-the-vagina
840 http://women.webmd.com/vaginal-dryness-causes-moisturizing-treatments
841 www.psychologytoday.com/blog/owning-pink/201104/15-crazy-things-about-vaginas
842 Ibid.
843 Ibid.
844 Ibid.
845 Ibid.
846 Ibid.
847 www.webmd.com/baby/features/6-embarrassing-pregnancy-symptoms
848 www.babycenter.com/0_uterine-rupture_1152337.bc
849 *Vaginas: An Owner's Manual* by Dr. Carol Livoti and Elizabeth Topp (Thunder's Mouth Press, 2004)
850 www.mayoclinic.com/health/vaginal-tears/PR00143&slide=5
851 www.mckinley.illinois.edu/handouts/toxic_shock_syndrome.html
852 Ibid.
853 http://ubykotex.com/real_answers/education/article?id=50289
854 *Blood, Bread and Roses: How Menstruation Created the World* by Judy Grahn (Beacon Press, 1993)
855 Ibid.
856 Ibid.
857 "Menstrual Blood Tapped as a Source of Stem Cells" by Steve Mitchell, MSNBC.com, November 30, 2007

858 www.gurl.com/clitoris-vagina-women-orgasm-sexual-pleasure/
859 Ibid.
860 www.examiner.com/sexual-health-in-national/clitoris-facts-to-make-you-all-tingly
861 www.msnbc.msn.com/id/14061671/ns/health-sexual_health/t/all-dressed-latex-dog-collars/#.TouEDXHW2jo
862 Ibid.
863 www.cbsnews.com/8301-504763_162-20013195-10391704.html
864 Ibid.
865 *Necrophilia: Forensic and Medico-Legal Aspects* by Anil Aggrawal (CRC Press, 2010)
866 www.cbsnews.com/8301-504763_162-20013195-10391704.html
867 Ibid.
868 Ibid.
869 *The Encyclopedia of Unusual Sex Practices* by Brenda Love (National Book Network, 1994)
870 www.cbsnews.com/8301-504763_162-20013195-10391704.html
871 Ibid.
872 *Handbook of Clinical Sexuality for Mental Health Professionals* by Stanley E. Althof (Taylor & Francis, 2010)
873 *The Strange Case of the Walking Corpse: A Chronicle of Medical Mysteries, Curious Remedies and Bizarre But True Healing Folklore* by Nancy Butcher (Avery, 2004)
874 *The Book of Kink: Sex Beyond the Missionary* by Eva Christina (Perigee, 2011)
875 *Exploring the Dimensions of Human Sexuality* by Jerrold S. Greenberg, Clint E. Bruess and Sarah C. Conklin (Jones & Bartlett Learning, 2010)
876 Ibid.
877 http://jezebel.com/354244/tyra-takes-on-sexual-squashing-fetish
878 *The Book of Kink: Sex Beyond the Missionary* by Eva Christina (Perigee, 2011)
879 Ibid.
880 Ibid.
881 Ibid.
882 Ibid.
883 *The Know-It-All: One Man's Humble Quest to Become the Smartest Man in the World* by A. J. Jacobs. Simon & Schuster, 2004)
884 *The Encyclopedia of Unusual Sex Practices* by Brenda Love (National Book Network, 1994)
885 *The Book of Kink: Sex Beyond the Missionary* by Eva Christina (Perigee, 2011)
886 Ibid.
887 *The Strange Case of the Walking Corpse: A Chronicle of Medical Mysteries, Curious Remedies and Bizarre But True Healing Folklore* by Nancy Butcher (Avery, 2004)

888 *The Book of Kink: Sex Beyond the Missionary* by Eva Christina (Perigee, 2011)
889 Ibid.; *The Quick-Reference Guide to Sexuality & Relationship Counseling* by Dr. Tim Clinton and Mark Laaser (Baker Books, 2010)
890 *The Book of Kink: Sex Beyond the Missionary* by Eva Christina (Perigee, 2011)
891 *The Strange Case of the Walking Corpse: A Chronicle of Medical Mysteries, Curious Remedies and Bizarre But True Healing Folklore* by Nancy Butcher (Avery, 2004)
892 Ibid.
893 Ibid.
894 Ibid.
895 www.menshealth.com/mhlists/penis_facts/Penis_Fact_6.php#ixzz1U0K4U95D
896 www.thefrisky.com/post/246-13-amazing-penis-facts/P1/
897 Ibid.
898 Ibid.
899 Ibid.
900 www.rollingstone.com/news/story/5938137/mr_big
901 www.msnbc.msn.com/id/13509704/ns/health-sexual_health/t/man--year-erection-awarded/#.Tn-kSk-GokA
902 Ibid.
903 Ibid.
904 www.menshealth.com/mhlists/penis_facts/Penis_Fact_6.php#ixzz1U0K4U95D
905 www.health.com/health/article/0,20429777,00.html
906 www.menshealth.com/mhlists/penis_facts/Penis_Fact_6.php#ixzz1U0K4U95D
907 Ibid.
908 Ibid.
909 Ibid.
910 www.netdoctor.co.uk/sex_relationships/facts/penissize.htm
911 www.menshealth.com/mhlists/penis_facts/Penis_Fact_6.php#ixzz1U0K4U95D
912 http://men.webmd.com/features/5-things-you-did-not-know-about-your-penis
913 Ibid.
914 Ibid.
915 Ibid.
916 Ibid.
917 www.psychologytoday.com/blog/homo-consumericus/200912/facts-and-myths-about-the-human-penis
918 Ibid.
919 Ibid.
920 Ibid.

921 http://guyism.com/humor/animals-with-the-largest-genitals.html
922 www.omg-facts.com
923 www.gurl.com/testicles-reproduction-male-anatomy-scrotum-testes-endocrine-system-facts/
924 Ibid.
925 www.cancer.ca/nutfacts
926 www.gurl.com/testicles-reproduction-male-anatomy-scrotum-testes-endocrine-system-facts/
927 Ibid.
928 www.gurl.com/testicles-reproduction-male-anatomy-scrotum-testes-endocrine-system-facts/
929 Ibid.
930 Ibid.
931 *Fertility Facts* by Kim Hahn and *Conceive Magazine* (Chronicle Books, 2008)
932 Ibid.
933 www.nlm.nih.gov/medlineplus/ency/article/003160.htm
934 www.gurl.com/testicles-reproduction-male-anatomy-scrotum-testes-endocrine-system-facts/
935 www.cancer.ca/nutfacts
936 www.cancer.org/Cancer/TesticularCancer/OverviewGuide/testicular-cancer-overview-key-statistics
937 Ibid.
938 Ibid.
939 Ibid.
940 www.davesdaily.com/interesting/interesting-useless.htm
941 www.cancer.ca/nutfacts
942 www.omg-facts.com/view/Facts/22047
943 www.cancer.ca/nutfacts
944 Ibid.
945 www.etymonline.com
946 www.cancer.ca/nutfacts
947 Ibid.
948 www.msnbc.msn.com/id/17951664/ns/health-sexual_health/t/many-cheat-thrill-more-stay-true-love/#.To0JuHHW2jo
949 www.afsp.org/index.cfm?page_id=04ECB949-C3D9-5FFA-DA9C65C381BAAEC0
950 http://marriage.about.com/cs/masturbation/f/masturbatfaq3.htm; *Sex for Dummies* by Ruth K. Westheimer and Pierre A. Lehu (John Wiley & Sons, 2006)
951 *Sex for Dummies* by Ruth K. Westheimer and Pierre A. Lehu (John Wiley & Sons, 2006)
952 www.cosmopolitan.com/advice/tips/great-female-survey/satisfying-sex-lives
953 Ibid.

954 Ibid.; www.askmen.com/specials/2010_great_male_survey/dating_results.html
955 www.cosmopolitan.com/advice/tips/great-female-survey/satisfying-sex-lives; www.askmen.com/specials/2010_great_male_survey/dating_results.html
956 www.cosmopolitan.com/advice/tips/great-female-survey/satisfying-sex-lives
957 http://familysafemedia.com/pornography_statistics.html#anchor4
958 *Global Prevalence and Incidence of Selected Curable Sexually Transmitted Infections: Overview and Estimates* (Geneva: World Health Organization, 2001)
959 http://std.about.com/od/overviewofstds/a/incubationper.htm
960 www.cdc.gov/std/chlamydia/STDFact-Chlamydia.htm
961 www.cdc.gov/std/chlamydia/default.htm
962 www.cdc.gov/std/chlamydia/STDFact-Chlamydia.htm
963 www.cdc.gov/std/gonorrhea/default.htm
964 www.cdc.gov/std/gonorrhea/STDFact-gonorrhea.htm
965 Ibid.
966 Ibid.
967 www.guardian.co.uk/theguardian/2003/jun/26/features11.g2
968 www.cdc.gov/hepatitis/PublicInfo.htm#whatIsHep
969 Ibid.
970 Ibid.
971 www.cdc.gov/std/herpes/STDFact-Herpes.htm
972 Ibid.
973 Ibid.
974 www.cdc.gov/std/HPV/STDFact-HPV.htm
975 Ibid.
976 www.cdc.gov/std/hiv/STDFact-STD-HIV.htm)
977 www.cdc.gov/std/trichomonas/STDFact-Trichomoniasis.htm)
978 www.cdc.gov/std/syphilis/STDFact-Syphilis.htm)
979 Ibid.
980 www.avert.org/std-statistics-america.htm
981 www.psychologytoday.com/blog/owning-pink/201104/15-crazy-things-about-vaginas
982 www.prlog.org/10510509-what-is-that-5-rare-stds.html
983 Ibid.
984 www.aarp.org/health/conditions-treatments/news-05-2011/seniors_sex_lives_are_up_and_so_are_std_cases.html
985 http://ezinearticles.com/?The-History-of-STDs&id=1565331
986 Ibid.
987 Ibid.
988 http://aids.about.com/cs/conditions/a/std.htm
989 http://aids.about.com/od/newlydiagnosed/a/hivtimeline.htm
990 Ibid.

991 http://aids.about.com/od/dataandstatistics/qt/worldstats.htm
992 www.cdc.gov/hiv/topics/perinatal/resources/factsheets/perinatal .htm
993 http://biotech.law.lsu.edu/Books/lbb/x590.htm
994 www.avert.org/std-statistics-america.htm
995 Ibid.
996 Ibid.
997 *Jet*, December 16, 1985, p. 24
998 *The Hite Report: A Nationwide Study of Female Sexuality* by Shere Hit (Seven Stories Press, 2003)
999 *On Kissing: From the Metaphysical to the Erotic* by Adrianna Blue (Wellington House, 1996)
1,000 Ibid.
1,001 www.thefrisky.com/photos/11-vagina-shockers/squirt-31010-g-jpg/
1,002 http://blogs.sfweekly.com/thesnitch/2009/05/masturbate-a-thon_champion_spe.php
1,003 http://news.softpedia.com/news/16-Reasons-Why-Sex-Is-Good-for-Your-Helath-67663.shtml

ABOUT THE AUTHOR

CARY McNEAL is an Emmy-winning TV writer/producer and author of two books, *1,001 Facts That Will Scare the S#*t Out Of You* (Adams) and *Crap I Bought on eBay* (Running Press). He lives in Atlanta, Georgia. Follow him on Twitter @carymcneal.